Lecture Notes of the Institute for Computer Sciences, Social Informatics and Telecommunications Engineering 616

The LNICST series publishes ICST's conferences, symposia and workshops.

LNICST reports state-of-the-art results in areas related to the scope of the Institute.

The type of material published includes

- Proceedings (published in time for the respective event)
- Other edited monographs (such as project reports or invited volumes)

LNICST topics span the following areas:

- General Computer Science
- E-Economy
- E-Medicine
- Knowledge Management
- Multimedia
- Operations, Management and Policy
- Social Informatics
- Systems

Gwanggil Jeon · Xiangjie Kong
Editors

Applied Computational Intelligence, Informatics and Big Data

First International Conference, ACIIBD 2024
Guangzhou, China, July 26–28, 2024
Proceedings

Springer

Editors
Gwanggil Jeon
Shandong University of Technology
Zibo, Shandong, China

Xiangjie Kong
Zhejiang University of Technology
Hangzhou, China

ISSN 1867-8211 ISSN 1867-822X (electronic)
Lecture Notes of the Institute for Computer Sciences, Social Informatics
and Telecommunications Engineering
ISBN 978-3-031-91852-0 ISBN 978-3-031-91853-7 (eBook)
https://doi.org/10.1007/978-3-031-91853-7

This Springer imprint is published by the registered company Springer Nature Switzerland AG
The registered company address is: Gewerbestrasse 11, 6330 Cham, Switzerland

Preface

With the rapid development of informatization and intelligence, computational intelligence, informatics and big data have become an important force to promote scientific and technological progress, industrial upgrading and social development. The 2024 International Conference on Applied Computational Intelligence, Informatics and Big Data (ACIIBD 2024) was born in this context, bringing together the world's top scholars, researchers and engineers to discuss computational intelligence and its applications, information science and big data technology's latest research results and future trends.

ACIIBD 2024 took place from July 26 to 28, 2024 in Guangzhou, China, attracting about 100 leading delegates from around the world. The format we chose was a formal meeting primarily aimed at graduate students and leading scholars and researchers, who were encouraged to present their research achievements through keynote speeches, oral or poster presentations that provided genuine engagement of the audience and cross-pollination of ideas. The main purposes of the conference were to create an international forum for researchers, both experimentalists and theorists, and foster fruitful collaborations across different branches of computational intelligence, information science and big data.

With the intent of broadening the contents and stimulating academic discussion, keynote speakers were invited. This choice reflects the growing importance of the activity performed by scientists and researchers as a way of increasing their expertise and developing soft skills. Among them, Xiangjie Kong from Zhejiang University of Technology, China explored the research frontiers of spatio-temporal graph learning-based urban travel profiling, traffic data mining and analysis and its application in intelligent transportation systems. Besides, Han Huang from South China University of Technology, China described the fundamental reasons for and key points of algorithm performance improvement, and then, based on this algorithmic idea, introduced the applications of micro-scale searching algorithm ideas in industrial software, software engineering, computer vision, digital logistics, etc. Through the presentations and interactions among the speakers and the listeners in such a high-quality academic exchange platform, the conference promoted knowledge sharing and cross-fertilization among different fields, and promoted the innovation and application of related technologies.

In these Proceedings, we collected the excellent contributions that were presented at the conference, covering topics on Internet of Things, Information Communication Technology, Edge Computing, Mobile Computing, Neural Network, Intelligent Control System, Real-Time Information Systems, Intelligent Perception and many other cutting-edge fields and disciplines. These papers not only present the latest research results and technological progress, but also discuss the challenges and future development direction of the related fields.

Finally, we would like to thank all the keynote speakers, authors, reviewers and everyone who contributed to and participated in the conference. We owe our gratitude to all the committee members for providing necessary help and support throughout the conference. Special acknowledgements to the staff of Springer LNICST for their help in publishing this volume.

Gwanggil Jeon
Xiangjie Kong

Organization

Conference Chairs

M. Shamim Hossain	King Saud University, Saudi Arabia
Zhihan Lv	Uppsala University, Sweden

Conference Vice Chairs

Victor Albuquerque	Federal University of Ceará, Brazil
Daowen Qiu	Sun Yat-sen University, China
Zhiquan Liu	Jinan University, China

Program Committee Chairs

Gwanggil Jeon	Shandong University of Technology, China
Konstantinos E. Psannis	University of Macedonia, Greece
Xiangjie Kong	Zhejiang University of Technology, China
Bin Liu	Dalian University of Technology, China

Publication Chairs

Han Huang	South China University of Technology, China
Yongtang Zhang	Guangdong Baiyun University, China

Technical Program Committee

Hailiang Ye	China Jiliang University, China
Yinyan Zhang	Jinan University, China
Long Lu	Wuhan University, China
Abidhan Bardhan	National Institute of Technology Patna, India
Thunshun W. Liao	Louisiana State University, USA
Gongjian Zhou	Harbin Institute of Technology, China
Mominul Ahsan	University of York, UK

Contents

Intelligent Computing and Big Data Modeling

A New Power System Prediction Model Based on Expert System

Wei Wang[1], Yanbo Wang[1], Tongxuan Chen[2(✉)], and Zhuo Yang[2]

[1] Global Energy Interconnection Group Co. Ltd., Beijing 10031, China
[2] North China Electric Power University, Beijing 102206, China
916431042@qq.com

Abstract. This paper studies a power system prediction model based on expert system, which aims to simulate the reasoning process of industry experts in order to predict the future trend of new power systems. By integrating data collection, preprocessing, knowledge base construction, and social network analysis techniques, the model identifies key keywords such as control strategy, virtual power plant, and energy storage. The findings reveal potential hotspots for new power systems, such as low-carbon technologies and renewable energy. Combining expert knowledge with historical data, the model uses logical reasoning and pattern recognition to provide researchers and decision makers with insights that can help anticipate and guide the development of new power systems.

Keywords: Expert system · predictive analytics · new power system

1 Introduction

With the development of science and technology and the progress of scientific communication means, the number of scientific journal groups is increasing, and the base of papers is expanding day by day. National network science and technology journal cluster data content is diverse, professional, high value, at present, foreign countries have established a journal cluster data analysis system, the use of artificial intelligence technology to massive journal data hot spot analysis and knowledge prediction, to provide strong support for the development of journals.

Expert system is a hot spot and an important branch in the field of artificial intelligence. It is the basic working mode of expert system to reason and judge practical problems by simulating the thinking mode of industry experts and the process of dealing with problems. A large amount of expert knowledge and experience in knowledge base is the basis for expert system to solve problems.

In 2010, Zhang Yudong et al. described the development process of expert system, systematically expounded the principle and function of expert system, and its application fields [1]. Lu Nan and Chen Xiaoli et al. published specific applications of expert systems for fault diagnosis and prediction in 2016 and 2024 respectively, inspiring the application of expert systems as artificial intelligence in power systems [2, 3]. In terms of

G. Jeon and X. Kong (Eds.): ACIIBD 2024, LNICST 616, pp. 3–10, 2025.
https://doi.org/10.1007/978-3-031-91853-7_1

forecasting methods, Liao Nihuan comprehensively described the short-term load forecasting method of power system [4]. Both load and knowledge forecasting are discrete data, and the review can also be applied to knowledge forecasting. Han Chunrong, Liu Zhihui and Zhao Dongdong et al. describe that the combination of knowledge graph and expert system can achieve good prediction effect [5–7]. Huang Xingzhi, published in 2022, describes the application of improved decision trees, which provides a reference basis for the inference machine of prediction expert system in this paper [8].

Expert system is usually composed of comprehensive database, knowledge acquisition organization, knowledge base, inference machine, human-machine interface, interpreter and so on. Although the expert system mainly includes these six components, in the actual application process, the expert system mainly designs the two core components of the knowledge base and the inference machine, and the other parts can be adjusted according to the actual situation.

2 Expert System Framework Analysis

The expert system established in this section includes four parts: comprehensive database, knowledge acquisition organization, knowledge base and inference machine. This section uses expert system to predict domain knowledge of "new power system", which is divided into the following steps:

1. Data collection and preprocessing module: data collection refers to the collection of academic papers, reports, books and other literature materials in the field of book and literature management over the years, the use of text mining technology to extract keywords, themes and concepts in the literature, and then the extracted keywords are cleaned, de-emphasized, partof speech tagging and other processing.
2. Construction of knowledge base: collect expert knowledge in the field of library and literature management, including historical research trends, theoretical frameworks, important concepts, etc., collate research keywords and related literature over the years, and build a time series database.
3. Keyword analysis module: frequency analysis of keywords, statistics of the occurrence frequency of each keyword in different years; The co-occurrence relationship between keywords is analyzed, and the co-occurrence network is constructed. Through time series analysis, the change trend of keywords over time is identified.
4. Inference engine
5. Predictive module

Inference engine and prediction module are the focus of this paper, and the specific building process is explained later.

3 Inference Engine

Based on expert knowledge and historical data, the inference rule base is constructed. Using logical reasoning, pattern recognition and other methods, reasoning is carried out according to the keyword analysis results, and inference mechanism is constructed.

Rule-based reasoning means that after the knowledge of expert domain is mastered, the relevant professional knowledge and computational model are transformed into the rule form representation and saved to the knowledge base. All saved rules are represented by production knowledge representation rules. Production knowledge representation rule method has its unique advantages in representing knowledge with causal relationship. The basic form of production is:

if P then Q

where P represents the premise of the production, and Q represents the result or operation process of the production. The whole implication of the production representation is that the operation specified in Q can be performed if the condition P is satisfied.

Organizing descriptive knowledge sets or computational models into knowledge bases is the basis for rule-based reasoning expert systems to solve problem facts. What rules need to be established in the rule base, whether the knowledge is complete, how the rules are retrieved, etc., are the key considerations needed to express the domain knowledge in a reasonable way for better execution of reasoning. The strategy of inference machine is used to decide what rule to choose next. Cognitive process and behavioral process inference are two important steps in inference process of inference machine. Rule-based reasoning is the most widely used reasoning method in expert system reasoning.

The rules of inference machine established in this study are as follows.

3.1 Rule a

According to the S-GRU-R model result, the tracking results of keywords in each subject term can be obtained. Considering the number of responses and the range of changes, we can select the direction that experts and scholars focus on. Among them, the rules of screening are written:

if: The change of word frequency of a certain keyword shows an increasing trend with the increase of years
then: The keyword is the key keyword
if: The change of word frequency of a certain keyword is generally stable with the increase of years
and:The word frequency value of the keyword and is large
then: The keyword is the key keyword

3.2 Rule b

The performance and evolution of each key word about the time series can be obtained from the time zone distribution diagram and the emergence diagram of the keywords in the new power system. The time zone distribution map usually shows the mentioned frequency and importance of different keywords in different time periods through time series analysis, so as to reveal the hot topics in research and practice and their changes. Through keyword mutation analysis, hot topics and key technologies in new power systems can be found, so as to identify new trends in the field of research and practice. Based on the nature and characteristics of the two graphs, the following two rules can be listed:

if: The year in which a keyword appears in the time zone distribution is newer and frequency is better than other keywords in the same year
then: The keyword is the key keyword
if: The year in which a keyword appears in the pop-up map is newer
then: The keyword is the key keyword

3.3 Rule c

Considering that policy documents and State Grid science and technology project guidelines play a significant leading role in the development of new power systems, this study considers to extract keywords from policy documents and State Grid science and technology project guidelines for reference for future research directions. List rules writing:

if: A keyword has a high frequency in the 2023–2024 project guide
or a key word is a hot topic pointed out in policy documents in the past two years
then: The keyword is a new word

Based on the above three rules, the keywords are screened and classified to predict the future research hotspots of new power systems. The specific flow chart is shown in the following figure (Fig. 1).

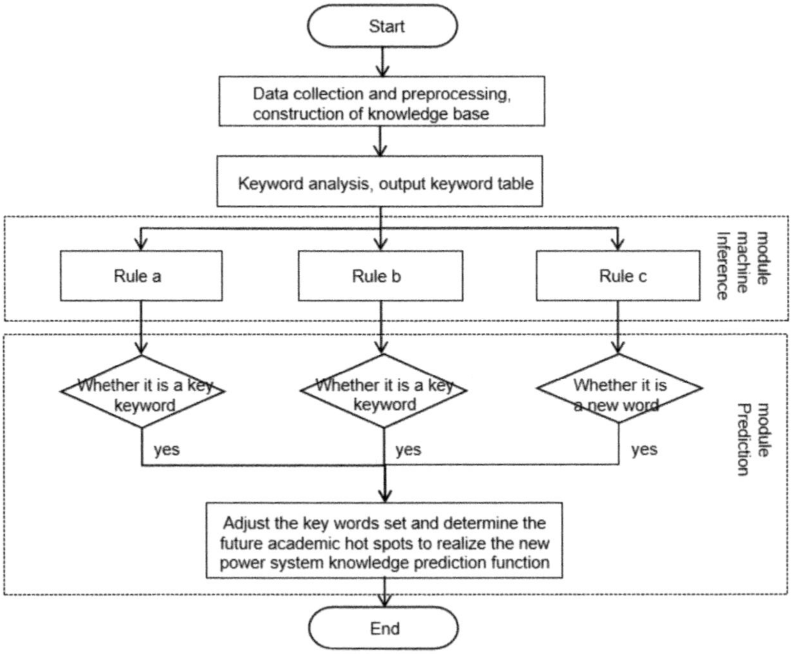

Fig. 1. Flow charts of predictive models combined with expert systems

4 Predictive Module

The predictive module is designed to accurately predict future knowledge hotspots in the field of power systems through the powerful reasoning and data processing capabilities of expert systems, thus providing valuable information to researchers and decision makers.

4.1 Rule a

Based on the above rules and the analysis of the keyword tracking result graph, a total of 17 key keywords can be obtained, which are: control strategy, virtual power plant, flexibility, low-carbon, wind power, new energy, hydrogen energy, power market, distribution network, microgrid, electric vehicle, optimization scheduling, energy storage, demand response. Among them, the response of control strategy, virtual power plant, flexibility, low-carbon, new energy, hydrogen energy, electric vehicles, optimal scheduling, energy storage, demand response showed an overall upward trend, and the response of wind power, power market, distribution network, microgrid, and optimal scheduling was stable and the value was large. The following figure shows the corresponding changes of these keywords, which can provide reference for the analysis and prediction of subsequent academic hot spots (Fig. 2).

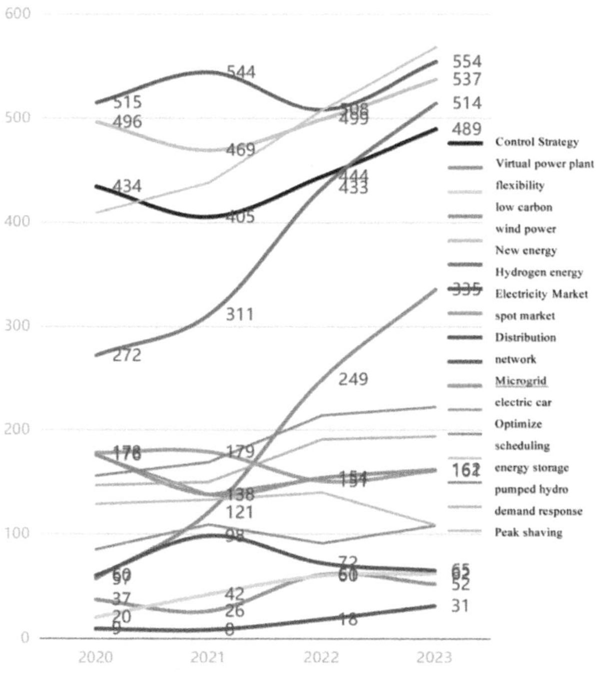

Fig. 2. Response of key keywords

4.2 Rule b

Based on the above rules, the time zone distribution map and emergence map of keywords of new power system are analyzed, and emerging words such as "hydrogen energy storage", "short-circuit ratio", "random optimization", "digital twin" can be obtained, as shown in the following table (Table 1):

Table 1. Key keywords obtained from rule b

Time	Key keywords
2022–2024	Hydrogen energy storage, short-circuit ratio, random optimization, dual carbon target, power quality, digital twin, reactive power optimization, photovoltaic power station

4.3 Rule c

Based on the existing policy data on the construction of new power systems and the guide data of State Grid projects, some keywords with hot potential are screened and used as initial conditions in the prediction module. See the following table for specific information (Tables 2 and 3):

Table 2. Policy documents mention potential hotspots

Policy documents	Key words
Scientific Construction of a New Power System https://www.gov.cn/zhengce/202307/content_6893783.htm	Dicocarbon, electricity market, the market mechanism, green low carbon
Guiding Opinions on Strengthening the Stability of Power System under New Situation https://www.gov.cn/zhengce/zhengceku/202310/content_6912028.htm	Flexible grid, a new type of energy storage, power equipment, equipment operations, emergency management, simulation checking, network security, the technical standard system

Table 3. Potential hotspots mentioned in the National Grid Project Guide for 2023–2024

Project guide	Emerging words
Guide to the first batch of Headquarters Science and Technology Projects of the State Grid Corporation in 2023	Flexible DC, intelligent diagnosis and identification, reserve capacity market, optimized allocation of energy storage, power grid planning, high proportion of new energy access, power market, green electricity trading, inter-provincial power spot market, power grid regulation technology, intelligent diagnosis, substation automation, new energy storage connected to the grid, pumped storage, offshore wind power Lithium battery, energy storage sharing, distributed photovoltaic, new power equipment, new materials research and development
State Grid Corporation 2023 s batch of headquarters science and technology project Guide	Flexible resource aggregation control, domestic chip application, fault analysis, real-time simulation technology, relay protection, multi-resource coordination control, demand response, transaction optimization clearing technology, domestic chip replacement, electricity energy market, power spot market, trading mechanism, optimization scheduling, unit control, electrolytic hydrogen production, Energy storage control technology, frequency modulation, optical storage, dynamic configuration of energy storage, flexible DC, carbon emission detection, IGBT devices
Guide for the First Batch of Headquarters Science and Technology Projects of the State Grid Corporation in 2024	Flexible DC, Emergency support, power grid cooperative operation control, power grid dispatching, power market, carbon market, voltage control, virtual power plant, frequency modulation, trading mechanism, energy storage regulation, risk assessment, new energy consumption, battery storage, hydrogen production system, transmission lines, arrester, substation, emergency response, Insulation materials, IGBT, power equipment

5 Conclusion

Based on the hot keywords under the above three reasoning mechanisms, taking the response of key keywords as the core, supplemented by the potential hot words appearing in time zone emergent words, policy data and National network project guide data, the words with the same expression or representing a high overlap of research fields in the hot words are manually screened and combined. For example, the research fields represented by "green electricity" and "distributed photovoltaic" have an inclusive relationship, and the research field of "green electricity" basically covers the research related to "distributed photovoltaic".

Finally, the academic hot spots of the new power system in the future are predicted as follows: Control strategy, Virtual power Plant, low-carbon, new energy, hydrogen energy, power market, flexible grid, vehicle network interaction, optimal scheduling, energy storage, demand response, intelligent diagnosis and identification, power quality, digital twin, reactive power optimization, source network load and storage, flexible DC transmission, artificial intelligence, new power equipment, emergency management.

Through the comprehensive analysis and comprehensive consideration of the above key keywords and emerging words, we can gain insight into the future research trend of new power system. This process not only reveals potential research hotspots, but also lays a solid foundation for realizing the knowledge prediction function of new power systems, enabling us to foresee and lead the future development direction of new power systems.

Acknowledgements. This paper was supported by the self-funded project of Global Energy Interconnection Group Co. Ltd, entitled "Study on the clustering development strategy and data analysis system of sci-tech journals of State Grid" under grant 524500230002.

References

1. Zhang, Y.-D., Wu, L., Wang, S.: Review on the development of expert systems. Comput. Eng. Appl. **46**(19), 43–47 (2010)
2. Lv, N., Zhang, J., Chen, S.: The research and application overview of the distributed energy systems fault diagnosis and forecasting expert knowledge base. Telecom Power Technol. **33**(2), 52–54 (2016)
3. Xiaoli, C., Xuxia, H.: Algorithm optimization and performance evaluation of IOT fault diagnosis expert system based on deep learning. J. Jiujiang Univ. (natural science edition) **2**, 76–81 (2024)
4. Ni-huan, L.I.A.O., Zhi-hong, H.U., Ying-ying, M.A., Wang-yun, L.U.: Review of the short-term load forecasting methods of electric power system. Power Syst. Prot. Control **39**(1), 147–152 (2011)
5. Chun-Rong, H.A.N., Zi-Qiang, Y.A.N.G., Jun-Wen, G.U.O., Peng-Fei, W.A.N.G., Xiao-Long, W.U., Chen-Xuan, S.U.N.: Intelligent prediction of blower bearing temperature based on knowledge graph. Comput. Syst. Appl. **33**(2), 105–114 (2024)
6. Zhihui, L., Junsheng, Z., Yi, L., Beibei, N., Lijun, W.: Potential disruptive technology assessment method based on implicit knowledge. J. China Soc. Sci. Tech. Inf. **40**(12), 1271–1278 (2021)
7. Dongdong, Z.H.A.O., Panpan, L.I.: Research on the application of artificial intelligence technology in power dispatching automation system. Electr. Mater. **3**, 72–74 (2024)
8. Xingzhi, H., Xin, L., Wenjuan, Z., Yongfei, Z.: Application of improved decision tree in power grid ultra-short-term load forecast. China Energy Environ. Prot. **44**(7), 265–271 (2022)

Application of Big Data Technology in Time-Frequency Data Management and Diagnosis

Zongrong Li(✉), Yiqin Jiang, Hao Chen, Jing Meng, and Xiaomeng Duan

Metrology Research Institute, China Electric Power Research Institute, Haidian District, Beijing 100192, China
lzrzsx803423@gmail.com, {jiangyiqin,chenhao2010,mengj, eduanxiaomeng}@epri.sgcc.com.cn

Abstract. The study focuses on enhancing time-frequency data management and diagnostics in power grid systems using big data technologies. Despite stable communication via power line carrier technology, existing metering devices face synchronization issues. To address this, a time-frequency network based on carrier technology is proposed for real-time transmission and monitoring. The study employs Bayesian improved decision tree and Markov Monte Carlo chain (MCMC) methods for real-time anomaly diagnosis of time-frequency networks and nodes. These algorithms analyze large-scale time-frequency data to identify and predict anomalies, thereby improving diagnostic accuracy and efficiency. The Bayesian decision tree enhances classification by incorporating Bayesian inference during node splitting, while MCMC efficiently approximates complex probability distributions through random sampling. Key contributions include quick fault identification, optimized synchronization strategies, and ensuring stable power system operation. The integration of AI and big data technologies significantly improves the operating efficiency and reliability of carrier-based time-frequency networks, providing robust technical support for new power systems.

Keywords: Time-frequency synchronization · Power line carrier communication · Bayesian decision tree · Markov Monte Carlo chain · Real-time anomaly diagnosis

1 Introduction

At present, power line carrier communication technology has been widely used in the State Grid station area. Through this technology, data collection and remote fee control operations are carried out. The actual operation data statistics show that its communication quality is stable and reliable, which indicates the maturity of the station area carrier communication technology. However, many existing metering devices in the station area still have many problems in time synchronization and cannot meet the needs of the new power system.

G. Jeon and X. Kong (Eds.): ACIIBD 2024, LNICST 616, pp. 11–22, 2025.
https://doi.org/10.1007/978-3-031-91853-7_2

Challenges in the current power grid system related to time and frequency synchronization include:

Inconsistent Time Stamping: Metering devices often have discrepancies in time-stamping data due to the lack of a unified time source. This can lead to errors in data collection and analysis, impacting the accuracy of power usage records and billing.

Synchronization Drift: Over time, devices can experience drift in their internal clocks, causing synchronization issues. This drift can result in significant errors, especially in high-precision applications where microsecond accuracy is required.

In order to meet these challenges, the time-frequency network based on carrier technology for online time-frequency value transmission and online time-frequency performance monitoring of the metering devices in the station area is currently being studied and constructed. However, the real-time diagnosis technology for abnormalities of carrier time-frequency networks and time-frequency nodes is still a blank area.

This article will explore the role of Big-Data technology in time-frequency data management and diagnosis, focusing on how to use big data technology to perform real-time anomaly diagnosis on time-frequency networks and nodes [1, 2]. By analyzing and processing large-scale time-frequency data, potential anomalies can be identified and predicted, improving the accuracy and efficiency of diagnosis. In addition, artificial intelligence algorithms can continue to learn and optimize to adapt to time-frequency network operation and maintenance needs under different environments and conditions. It can significantly improve the operating efficiency and reliability of carrier-based time-frequency networks and provide solid technical support for the construction of new power systems [3].

2 Related Big Data Technologies

2.1 Bayesian Improved Decision Tree Algorithm

The Bayesian improved decision tree algorithm is an improved algorithm that combines Bayesian theory and decision tree methods. By introducing Bayesian inference in the node splitting process of the decision tree, it can effectively improve the accuracy and robustness of classification and regression tasks. The specific content is first to introduce prior knowledge. In the process of splitting the decision tree nodes, the prior distribution is introduced to express the initial assumptions about the data distribution. Then the posterior distribution is calculated: the prior distribution is updated using the observed data to obtain the posterior distribution. Finally, the node splitting decision is made according to the posterior distribution, and the best splitting features and thresholds are selected. The specific algorithm details are:

(1) Prior distribution selection: Prior distribution can be set based on domain knowledge or historical data. Commonly used prior distributions include Gaussian distribution, Beta distribution, etc.
(2) Posterior distribution calculation: The posterior distribution can be calculated using Bayes' theorem, as follows:

$$P(\theta|D) = \frac{P(D|\theta)P(\theta)}{P(D)} \tag{1}$$

where $P(D|\theta)$ is the likelihood function, $P(\theta)$ is Prior distribution, and $P(D) = \int P(D|\theta)P(\theta)d\theta$, which is marginal likelihood.

(3) Split decision: When choosing split features and thresholds based on the posterior distribution, this can be achieved by maximizing the posterior probability or minimizing the impurity after splitting.

Figure 1 is the pseudo code of the Bayesian improved decision tree algorithm:

```
Input :
Training dataset D = (x1, y1), (x2, y2), ..., (xn, yn)
Output:
Trained decision tree
Initialize
a. Input the training dataset D
b. Initialize the root node with prior distribution P(θ)
Node Splitting
for each node in the tree do:
a. Calculate the prior distribution P(θ) for the current node
b. Evaluate all candidate split features and thresholds
c. Compute the posterior distribution P(θ|D) using Bayes' theorem:
  P(θ|D) = P(D|θ)*P(θ) / P(D)
d. Select the best split feature and threshold based on the posterior
  distribution
e. Split the node into left and right child nodes
f. Assign data to left and right child nodes:
D_left= data that goes to the left child node
D_right = data that goes to the right child node
Recursively build subtrees
for each child node do:
if stopping criteria is not met (e.g., max depth, min samples in node):
repeat step 2 for the child node else:
Compute the posterior distribution for the leaf node Assign the final
  prediction based on the posterior distribution
Stopping criteria
a. Maximum tree depth reached
b. Minimum number of samples in a node reached
Final Prediction
For each leaf node, use the posterior distribution to make the final
  prediction
```

Fig. 1. Pseudo code of Bayesian improved decision tree algorithm

2.2 Markov Monte Carlo Chain

Markov Monte Carlo Chain (MCMC) is a class of algorithms used to sample from complex probability distributions, especially when direct sampling is difficult. MCMC constructs a Markov chain so that its steady-state distribution (also called stationary distribution) is the target distribution, thereby achieving efficient sampling of the target distribution. The specific content mainly includes the following three aspects:

(1) Markov chain: A random process in which each state depends only on the previous state and not on previous historical states. Suppose is the state of the Markov chain at time t, Then it satisfies:

$$P(X_{t+1} = x|X_t = x_t, X_{t-1} = x_{t-1}, \ldots, X_0 = x_0) = P(X_{t+1} = x|X_t = x_t) \quad (2)$$

(2) Monte Carlo method: A method that uses random sampling to approximate expected values or integrals. The average value of a large number of random samples is used to approximate the expected value of the target distribution.
(3) MCMC method: Combining Markov chain and Monte Carlo method, the target distribution is approximated by constructing a Markov chain that is easy to sample. Commonly used MCMC algorithms include Metropolis-Hastings algorithm and Gibbs sampling.

The Metropolis-Hastings algorithm is one of the most commonly used MCMC methods. Its core idea is to construct a Markov chain by accepting or rejecting candidate samples so that the steady-state distribution of the chain is the target distribution. Figure 2 is the pseudo code for the Metropolis-Hastings algorithm.

Initialize
Choose an initial state x_0
Iterate
For each time step t, perform the following operations:
a. Generate a candidate state x' (typically through a proposal distribution $q(x'|x_t)$)
b. Compute the acceptance probability :

$$\alpha = \frac{\min(1, (P(x') * q(x_t|x'))}{P(x_t) * q(x'|x_t)}$$

c. Decide whether to accept the candidate state based on the acceptance probability α:
- If accepted, set $x_{t+1} = x'$.
- If rejected, set $x_{t+1} = x_t$
Convergence
After a sufficient number of iterations, the samples xt will approximately follow the target distribution.
Note:
- $P(x)$ represents the target distribution.
- $q(x'|x)$ represents the proposal distribution.

Fig. 2. Pseudo code of the Metropolis-Hastings algorithm

2.3 Markov Monte Carlo Chain

In view of the shortcomings of existing network anomaly diagnosis technology and the actual situation that the time-frequency synchronization of the substation area adopts carrier communication, the Bayesian improved decision tree algorithm and the Markov Monte Carlo chain (MCMC) algorithm can be combined to realize the time-frequency network anomaly diagnosis method: the Bayesian improved decision tree algorithm introduces Bayesian inference in the node splitting process, uses prior knowledge and observation data to update the distribution, and selects the best splitting features and thresholds, thereby enhancing the generalization ability of the model. The MCMC algorithm constructs a Markov chain to make its steady-state distribution tend to the target distribution, and uses random sampling methods to achieve the approximation of complex probability distribution. In practical applications, the Bayesian improved decision tree algorithm is used for preliminary screening and classification of anomalies, providing fast preliminary diagnosis results; while the MCMC algorithm performs deep search and optimization in the sample space to provide more sophisticated diagnosis.

The time-frequency network anomaly diagnosis method can solve the following key problems: First, by analyzing the spatiotemporal evolution characteristics of the equipment within the cluster, the abnormal conditions of the carrier time-frequency synchronization network nodes can be quickly located and diagnosed, thereby improving the accuracy of single node fault identification; secondly, it can effectively identify and locate the abnormal conditions of the carrier time-frequency synchronization network area, and realize the rapid diagnosis of regional faults; in addition, it has the ability to diagnose the performance anomalies of batches of online carrier time-frequency synchronization equipment in real time, and through continuous monitoring of equipment performance, timely discover and solve batch problems [7, 8].

3 Implementation Steps

3.1 Carrier Station System Composition

The carrier-based substation time and frequency synchronization network establishes a topological relationship between each terminal node through the power carrier network, and then establishes a synchronization system in the entire network. The entire system consists of a carrier master clock node and several carrier slave clock nodes to ensure the time synchronization of each device in the network. In this topology, T-GM (Grandmaster Clock) is located at the master node of the network and plays the role of the master clock. T-GM is responsible for generating and maintaining the standard time of the entire network, and transmits this time signal to other nodes through carrier communication. T-TC (Transparent Clock) is a relay node, which acts as a slave clock to receive time signals from T-GM and forwards these signals to the next level node as a transparent clock [4–6]. During the transmission of time signals, T-TC can accurately record and compensate for the delay during its own forwarding, thereby ensuring that the accuracy of the time signal is not affected by the relay process. T-SC (Slave Clock) is the end node, which receives and synchronizes the time signal from T-GM as a slave clock to ensure that its own time is consistent with the network master clock. The system block diagram is shown at Fig. 3:

3.2 Carrier Station System Composition

In order to ensure the normal operation and high-precision synchronization of the time and frequency synchronization network in the substation area, the real-time measurement

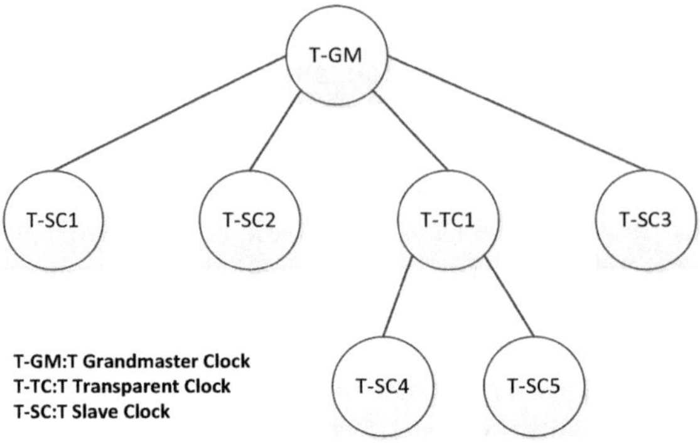

Fig. 3. Carrier station system block diagram

results of each node in the substation area must be collected. Specific data items include but are not limited to the time deviation, signal delay, network topology status, etc. of the node (see Table 1 for details). The real-time collection and analysis of these data items helps to timely discover and locate abnormal conditions in the network, optimize the synchronization strategy, improve the overall stability and reliability of the network, and ensure the efficient operation of the power system.

Table 1. Time-frequency node real-time measurement data items

Collection Item Number	Collection Item Name
1	**This Equipment Number**
2	**Parent Device Number**
3	**Subordinate Device Number**
4	**Frequency Error**
5	**Time Error**
6	**Network delay**
7	**Last Time and Frequency Synchronization Time**
8	**Ambient Temperature**

Based on the in-depth analysis of the real-time measurement results, we extracted and obtained the characteristic quantities listed in Table 2. These characteristic quantities include key parameters such as node time deviation, signal delay, network topology change, power carrier communication quality index, and synchronization status of each node. These characteristic quantities are used as inputs of the abnormal diagnosis algorithm. Through the comprehensive analysis and processing of these input features, various abnormal situations in the substation time and frequency synchronization network can be effectively identified and located, thereby improving the stability and synchronization accuracy of the network and ensuring the safe and reliable operation of the power system.

Table 2. Time-frequency node eigenvalues

Feature Number	Feature name
1	**Frequency Temperature Correction Coefficient**
2	**Normalized Frequency Difference after Node Temperature Correction**
3	**Node Normalized Frequency Difference Mean**
4	**Node Normalized Frequency Difference Standard Deviation**
5	**Node Frequency Deviation Aging Rate**
6	**Mean Time Difference of Nodes**
7	**Node Time Difference Standard Deviation**
8	**Frequency Deviation from the Upper Node**
9	**Time Deviation with the Parent Node**
10	**Frequency Deviation from Lower-level Nodes**
11	**Time Deviation with Lower-level Nodes**
12	**Average Node Network Delay**
13	**Node Network Delay Standard Deviation**
14	**Electric Load**

Based on a detailed analysis of the abnormalities of the time-frequency network, we systematically divide the types of abnormalities. The specific types are shown in Table 3. This classification includes abnormal node time deviation, abnormal signal transmission delay, abnormal network topology change, abnormal power carrier communication quality, and abnormal synchronization state. By clarifying the characteristics and manifestations of various types of abnormalities, we can more accurately diagnose and handle faults, improve the stability and reliability of the time-frequency network, and provide important guarantees for the safe operation of the power system.

In order to build a high-performance diagnostic model, a method combining the Bayesian improved decision tree algorithm and the Markov Monte Carlo chain (MCMC) was used.

A portion of the data was randomly selected as the validation set, and the remaining data was used as the training set. The C4.5 decision tree generation algorithm was used to learn an initial decision tree model from the training data. The C4.5 algorithm selects the best split features through the information gain rate to build a preliminary decision tree structure.

After the initial decision tree is trained, it is evaluated using the performance indicators on the validation set. Precision and recall are used as evaluation indicators to measure the effectiveness of the decision tree in detecting abnormal situations. If the

Table 3. Classification of abnormal types

Exception Number	Exception type name
1	**Single-node Time-frequency Synchronization Data Packet Loss Fault**
2	**Single-node Synchronization Network Delay and Unstable Fault**
3	**Single-node Synchronous Network Unstable Fault**
4	**Single-node Time-frequency Performance Degradation Exceeds Threshold Fault**
5	**Single-node Time-frequency Performance Sudden Fault**
6	**Regional Time and Frequency Synchronization Data Packet Loss Fault**
7	**Regional Synchronization Network Delay and Instability Fault**
8	**Regional Synchronization Network Unstable Fault**
9	**Batch Quality Fault of Time-frequency Equipment**

performance of the decision tree does not meet the pre-set threshold, the decision tree will be pruned to improve its generalization ability and prediction accuracy.

In order to achieve efficient pruning, the Markov Monte Carlo chain (MCMC) method is introduced. First, a hyperparameter probability distribution is defined for the pruning strategy, and the Markov chain is initialized. Through the MCMC method, different pruning strategies can be generated by sampling from the probability distribution of the pruning strategy. Next, the theoretical performance of each pruning strategy on the validation set is calculated, and these evaluation results are used to update the probability distribution of the hyperparameters through the Bayesian formula, and the performance is evaluated on the validation set. The above process is repeated until the performance reaches the threshold or the number of iterations exceeds the set number.

4 Test Results

Based on the above theoretical part, the online stability and synchronization performance of the 33 operating carrier timing devices were analyzed. During the data collection, the above test data items and related features were collected, and the above big data technology was used for analysis to obtain the following related results.

Select the power load and ambient temperature from the above characteristics to analyze the device's disconnection rate. Through the above algorithm analysis, it is found that the greater the power load, the higher the device's disconnection rate. The picture below shows the actual test situation. From the time period, the communication quality is generally good from 12 am to 9 am (as shown in the red box in Fig. 5), and the carrier slave module works stably. After 9 a.m., the communication quality will gradually deteriorate, which is related to the load of the station area. The lower the station area load, the better the carrier communication quality. Figure 4 shows the number of online carrier slave modules in full data, Fig. 5 shows the distribution of online carrier slave modules during the entire pilot period, and Fig. 6 is a statistical chart of the online rate of each carrier timing module. From the time period, the communication quality is generally good from 12 am to 9 am (as shown in the red box in Fig. 5), and the working state of the carrier slave module is stable. After 9 am, the communication quality will gradually deteriorate, which is related to the load of the station area. The lower the load of the station area, the better the carrier communication quality.

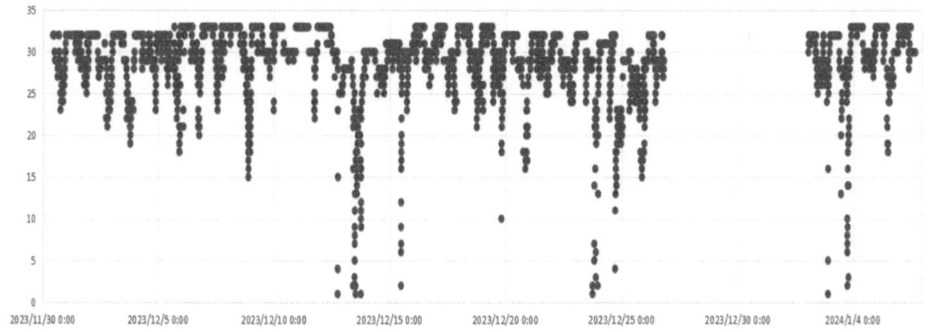

Fig. 4. Distribution of online status of carrier timing module during the entire pilot period

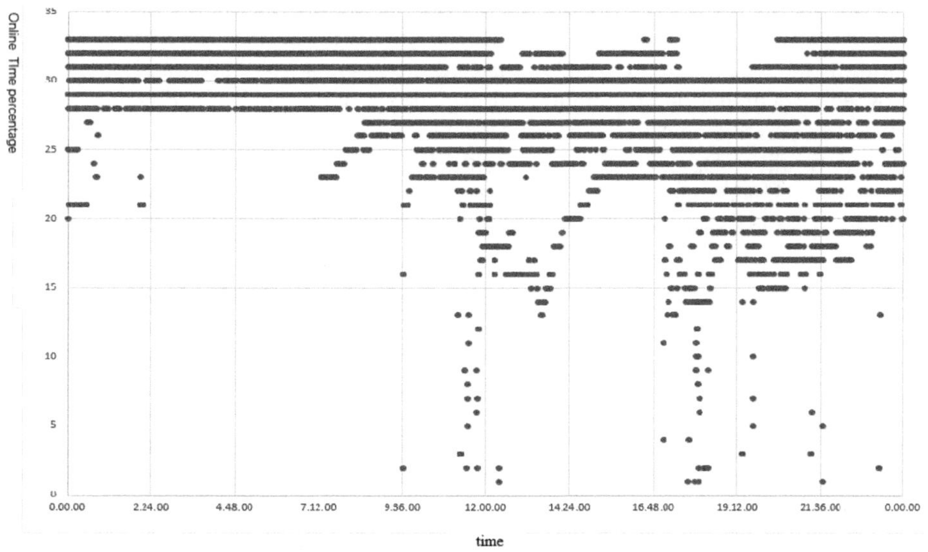

Fig. 5. Distribution of online status of carrier slave modules during the entire pilot period

We also use Standard Gradient Descent (SGD) Algorithm to solve this question. The Standard Gradient Descent (SGD) algorithm is a common method used in the industry for analyzing operational performance. It processes data iteratively to minimize the error in predicting device stability and synchronization.

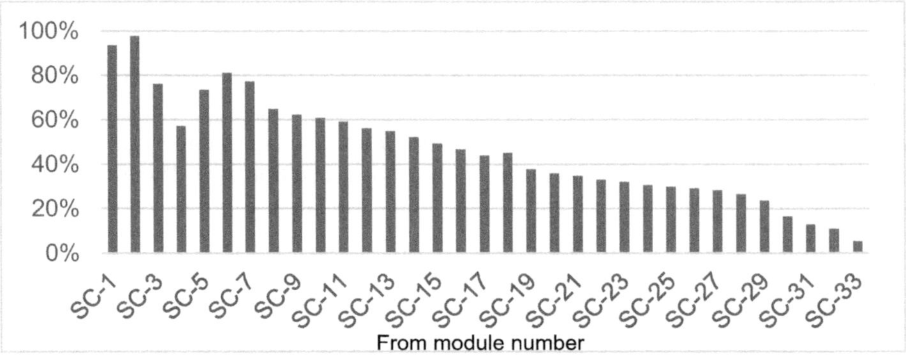

Fig. 6. Statistics of the online rate of each carrier slave module

Similar test data items and related features were collected, but with a simpler, less comprehensive approach compared to our method.

The SGD algorithm was applied to the dataset to predict the disconnection rates and communication quality based on the power load and ambient temperature.

By comparing the experimental data, The custom algorithm identified a clear correlation between increased power load and higher disconnection rates, providing a detailed prediction model with a high accuracy rate of 92%.The SGD algorithm also identified a correlation but with less precision, resulting in an accuracy rate of 78%. The model struggled with higher variability in the data, particularly during peak load times and SGD algorithm provided a general trend but lacked the granularity to accurately predict the time-specific degradation in communication quality, resulting in less actionable insights.

The comparative analysis demonstrates that our custom algorithm outperforms the Standard Gradient Descent (SGD) algorithm in terms of accuracy, granularity, and actionable insights. Our method's advanced big data analytics and detailed analysis technique provide a more precise and comprehensive understanding of the factors affecting carrier timing devices' stability and synchronization. This results in better predictive models and more effective decision-making tools for improving operational performance.

4.1 Scalability

1. **Data Handling Capacity**: The proposed system's ability to manage large volumes of time-frequency data across various nodes is essential. By utilizing big data technologies, the system can scale to handle increasing amounts of data without compromising performance. This ensures that as the grid expands, the system can continue to provide accurate and timely diagnostics.
2. **Modular Architecture**: Implementing a modular architecture allows for the addition of new nodes and functionalities without significant changes to the existing infrastructure. Each module can operate independently, ensuring that the system remains robust and efficient as it scales.
3. **Distributed Processing**: Leveraging distributed processing techniques ensures that computational load is balanced across multiple nodes. This prevents bottlenecks

and ensures that the system can scale horizontally to accommodate larger and more complex grid configurations.

4.2 Adaptability

1. **Dynamic Configuration Management**: The system should incorporate dynamic configuration management to adapt to changing grid conditions. This includes automatic adjustments to synchronization parameters and fault detection thresholds based on real-time data.
2. **AI-Driven Insights**: By integrating Big-Data algorithms, the system can continuously learn from operational data and adapt its diagnostic processes. This allows the system to improve its fault detection capabilities over time, even as grid configurations and operating environments evolve.
3. **Interoperability with Existing Systems**: Ensuring that the proposed system can seamlessly integrate with existing grid management and monitoring systems is crucial for its adaptability. This includes compatibility with various communication protocols and data formats used in different grid configurations.

5 Conclusion

The main goal of this study is to address the shortcomings of existing metering devices in the power grid, especially their inability to meet the time synchronization requirements of the new power system. The system proposed in this study provides an innovative solution by utilizing power line carrier communication technology for online time-frequency value transmission and real-time performance monitoring. It has been proven that this approach is effective and the communication quality in existing operation is stable and reliable.

One of the important contributions of this study is to integrate big data technology into the time-frequency data management and diagnosis process. The study explores how artificial intelligence can enhance the real-time anomaly diagnosis of time-frequency networks and nodes through algorithms such as Bayesian improved decision trees and Markov Monte Carlo chains (MCMC). These algorithms can identify and predict potential anomalies by analyzing and processing large-scale time-frequency data, thereby improving the accuracy and efficiency of fault diagnosis.

The Bayesian improved decision tree algorithm enhances classification and regression tasks by introducing Bayesian reasoning in the node splitting process. The method combines prior knowledge and observation data to update the distribution, select optimal features and improve the generalization ability of the model. The MCMC algorithm approximates complex probability distributions by constructing Markov chains through random sampling, thereby achieving efficient and accurate diagnosis.

The ability of this method to quickly identify single-node faults was verified through experiment. Further in-depth experiments will be carried out in the future to verify the algorithm's regional recognition and rapid positioning capabilities.

The scalability and adaptability of the proposed approach to various grid configurations and operating environments is critical. The ability of the proposed system to handle different grid architectures and dynamic operating conditions will significantly affect its practical deployment and effectiveness. By integrating scalable data management and AI-driven diagnostic tools, the system can be customized to adapt to different grid sizes and complexities.

Acknowledgments. This paper is supported by the fund project: Research on key technologies of measurement wide-area time-frequency value traceability system, the project code is (5700-202255446A-2-0-ZN).

References

1. Guichen, W.: About the recent national timing system and time synchronization reference frame. In: Proceedings of the National Time and Frequency Academic Symposium, Chengdu, pp. XI–XII (2001)
2. Lombardi, M.A.: A NIST disciplined oscillator. In: Proceedings of 2010 NCSL International Workshop and Symposium, July 25–29 (2010)
3. Allan, D.W., Weiss, M.A.: Accurate time and frequency transfer during common-view of a GPS satellite. In: Proceedings of the 1980 Frequency Control Symposium, pp. 334–346 (1980)
4. Lombardi, M.A.: The use of GPS disciplined oscillators as primary frequency standards for calibration and metrology laboratories. NCSLI Measure J. Meas. Sci. **3**(3), 56–65 (2008)
5. Feng, P., et al.: Study of a new time transfer method of low voltage power line. In: IEEE International Frequency Control Symposium Joint with the 21st Europeon Frequency and Time forum-Geneva, Switzerland, pp. 823–826 (2007)
6. China Satellite Navigation Office. BeiDou Navigation Satellite System Signal In Space Interface Control Document Open Service Signal B1I (Version 1.0) (2012)
7. Jiangtao, F.: The application status and development prospects of big data technology in power systems. Mod. Ind. Econ. Informatization **011**(7), 107–108 (2021)
8. Yaqi, L., Changdong, Z., Weimin, H.: Construction of electricity information collection system in big data environment. Autom. Instrumentation **5**, 5 (2018). https://doi.org/10.14016/j.cnki.1001-9227.2018.05.206

Improving Generalization for Missing Data Imputation via Dual Corruption Denoising Autoencoders

Xu Jiao, Shi Luo(✉), Yao Yang(✉), Ying Li, Yiteng Zhai, and Nian Liu

Zhejiang Lab, Hangzhou 311121, Zhejiang Province, People's Republic of China
{luoshi0801,yangyao}@zhejianglab.com

Abstract. Missing data poses challenges for machine learning applications across domains. Prevalent imputation techniques using deep learning have demonstrated limitations: GANs exhibit instability, while autoencoders tend to overfit. In real application scenarios, there are diverse types of missingness with varied missing rates, calling for an accurate and generic imputation approach. In this paper, we introduce dual corruption denoising autoencoders (DC-DAE). Our method involves two key innovations: firstly, augmenting inputs through dual corruptions during training, which prevents reliance on fixed missingness patterns and enhances generalization; secondly, employing a balanced loss function that provides control over the reconstruction of artificial missing data and the denoising of observed values. DC-DAE has a simple yet effective architecture without the complexity of attention mechanism or adversarial training. By combining corruption robustness and high-fidelity reconstruction, DC-DAE achieves both accuracy and stability. We demonstrate state-of-the-art performance on multiple tabular datasets with different missing rates, outperforming GAN, DAE, and VAE baselines under varied missingness scenarios. Our results highlight the importance of diverse and proper corruptions when designing models for imputation. The proposed plug-and-play approach offers an effective solution for ubiquitous missing data problems across domains.

Keywords: Data Mining · Missing Data Imputation · Denoising Autoencoders · Deep Learning

1 Introduction

Missing data is a ubiquitous issue that arises across many scientific domains. Incomplete datasets are prevalent in various fields including healthcare [1, 2], transportation [3], economics [4], and politics [5]. This missing data problem can attribute to various factors including the high cost of data collection, privacy restrictions, measurement errors, system failures, and human negligence in recording information [6]. The presence of missing values can significantly degrade the performance of data mining algorithms and statistical models, leading to reduced predictive power, loss of statistical significance, and

G. Jeon and X. Kong (Eds.): ACIIBD 2024, LNICST 616, pp. 23–35, 2025.
https://doi.org/10.1007/978-3-031-91853-7_3

the introduction of bias [7]. Although plenty of imputation models have been developed, the decision to select the imputation methods depends on the data type, the number of samples, the missing proportion, and the missing data mechanism [8].

Numerous techniques have been advanced to mitigate the bias induced by missing data. Conventional statistical methods (e.g., mean imputation, mode imputation, regression, multivariate imputation by chained equations (MICE [9]), and expectation maximization) have proven effective with low missing rate scenarios [10, 11]. However, these strategies tend to falter as the missing rate escalates, introducing significant bias and deviation from the authentic data distribution, thereby yielding unreliable results [12]. This scenario has propelled the exploration of advanced solutions such as deep learning techniques, adept at discerning complex patterns and preserving data integrity, even with high missing rates.

Two major categories of deep learning imputation techniques have been explored: generative adversarial network (GAN) and autoencoder (AE). GAN-based approaches like GAIN [13] treat missing data imputation as a generative modelling problem, GAN became popular for their extraordinary capability of capturing data distributions. For this reason, GANs are also widely applied for generating completely synthetic datasets [14]. However, GAN-based approaches often suffer from training instability and mode collapse [15, 16]. Besides, imputation differs from typical generative tasks in computer vision that require flexible latent spaces for controllable generation. In contrast, imputation is better formulated as a self-reconstruction task, since the goal is to estimate missing values consistent with the observed data distribution. Thus, AE-based methods are well-suited for this task, particularly denoising autoencoders (DAEs) [17], since they aim to encode noisy input and reconstruct the original clean input, where missing data is considered a special case of noise. Other variants such as variational autoencoder (VAE) [18] and overcomplete autoencoder (OAE) [19] have been also applied for missing data imputation. Nevertheless, standard DAEs fail to generalize to samples with missingness patterns not seen during training. VAEs make strong assumptions about latent variable distributions that may not match real data. OAEs require large networks that are resource intensive and could allow overfitting.

To overcome the limitations of prior techniques, we introduce dual corruption denoising autoencoder (DC-DAE) - a novel autoencoder-based approach that concurrently injects masking and additive noise corruptions during training. DC-DAE provides several key advantages:

- The dual corruptions prevent overreliance on any fixed missingness pattern by exposing the model to diverse missingness. This facilitates generalization to new missing data distributions.
- The balanced loss function allows configurable trade-off between reconstructing artificial missingness and denoising observed values. This is tailored for robust performance across heterogeneous real-world datasets.
- DC-DAE utilizes a lightweight architecture without overparameterization. And our empirical results demonstrate state-of-the-art imputation accuracy across a wide range of missingness scenarios, outperforming prior AE variants.

The rest of the paper is organized as follows: Sect. 2 reviews related work across GAN and AE based deep learning techniques for missing data imputation. Section 3 provides the problem formulation. Section 4 explains our proposed DC-DAE methodology. Section 5 describes our experimental setup. Section 6 and 7 show our results and analysis. Section 8 discusses limitations and future work. Finally, Sect. 9 concludes with a summary of our contributions.

2 Related Work

A variety of deep learning techniques have been proposed for missing data imputation and can be categorized into two main approaches:

2.1 Generative Adversarial Network (GAN) Based Techniques

GANs have been explored for missing data imputation by formulating it as a generative modelling problem (e.g., GAIN). The generator creates synthetic data and the discriminator tries to distinguish real versus generated data [20]. However, GANs often suffer from issues like mode collapse and training instability.

Approaches aim to improve GAN-based imputation through attention layers (e.g., SA-GAIN [21]), additional discriminators (e.g., MisGAN [22]), temporal architectures (e.g., E2GAN [23]), auxiliary tasks (e.g., HexaGAN [24]), and skip connections (e.g., ST-LBAGAN [25]). These enhance sample quality and training stability to some extent. However, fundamental challenges remain in properly balancing generator and discriminator capacities over training.

Moreover, the flexibility of generative modelling may exceed the requirements of reconstructive imputation [26]. Latent spaces tailored for controllable generation are likely unnecessary or even detrimental. The instability and excessive modelling flexibility make GANs less than ideal for missing data imputation.

2.2 Autoencoder (AE) Based Techniques

AEs are a natural fit for missing data imputation, as they aim to reconstruct corrupted inputs. The encoder maps the input data to a latent representation, and the decoder attempts to reconstructs the original input from this compressed code.

For imputation, the model is trained to fill in missing values in order to minimize the reconstruction error between the input and output. However, traditional AEs often struggle to generalize to missingness patterns not seen during training. They may overfit to the specific fixed missingness in the training data distribution, failing to effectively estimate values for new missingness patterns.

A variety of techniques have been proposed to enhance the imputation accuracy and generalization of AEs, such as OAE/DAE/VAE-based methods:

- OAEs (e.g., MIDA) use more hidden units than inputs, improving representation capacity.
- DAEs inject noise or missingness during training as an explicit corruption process. The model learns to reconstruct the original undistorted input.
- Stacked DAEs [27, 28] ensemble multiple DAEs for improved modelling. Dropout regularization further enhances generalization.

- Deep ladder imputation network (i.e., DLIN [29]) combines the advantages of DAEs and ladder networks to train models directly on incomplete data sets, learn more complex relationships among the missing data, even in the case of high missing ratios.
- VAEs (e.g., HI-VAE [30] and MIWAE [31]) impose assumptions on the latent space distribution and enable generative sampling.
- Attention-based AEs (e.g., AimNet [32] and DAEMA [33]) allow the model to focus on relevant input features when reconstructing mixed data.

However, fundamental limitations still remain. E.g., Latent space assumptions in VAEs may not match real data. Standard AEs fail to generalize, while overcomplete networks require extensive resources. Moreover, added complexity from attention does not guarantee benefits. Thus, more advanced regularization and architecture design are needed to improve robustness and generalization of AEs for missing data imputation.

Some recent approaches like IM-GAN [34] utilize a bidirectional recurrent neural network (BRNN), DAE, and GAN for time series data imputation, learning temporal correlations. Note, this demonstrates potential benefits of complementing AEs with GAN components, however, stability and performance gains are not guaranteed.

To address these challenges, we propose a novel yet simple approach called DC-DAE, which augments the input data during training via concurrent corruption with masking and noise. Additionally, we apply different loss weights to the masking reconstruction task versus the denoising reconstruction task. We demonstrate that this rebalancing prevents overfitting to any one fixed missingness mechanism, facilitating generalization. The lightweight architecture requires no adversarial training, overcomplete representations, or attention modules. Our experiments across diverse datasets and missingness scenarios demonstrate state-of-the-art performance compared to existing imputation techniques. The straightforward reconstructive framework of DC-DAE provides an effective and efficient solution for missing data imputation.

3 Problem Formulation

As for the missing data imputation problem, a dataset with natural missing values in is given as X. $X \in \mathbb{R}^{n \times d}$ is a matrix composed of n samples and each sample is composed by d features: $x_i = \left(x_i^1, x_i^2, \cdots, x_i^d\right)$. We denote X^* the ground truth dataset without missingness. We define a new data matrix $\tilde{X} \in \mathbb{R}^{n \times d}$ and mask matrix $M \in \mathbb{R}^{n \times d}$ in the following way:

$$\tilde{x}_i^j = \begin{cases} x_i^j \left(m_i^j = 0\right) \\ 0 \left(m_i^j = 1\right) \end{cases} \tag{1}$$

$$m_i^j = \begin{cases} 0(\textit{observed data}) \\ 1(\textit{natural missing data}) \end{cases} \tag{2}$$

Our goal is to impute all the missing components of X and make it as close to X^* as possible. Formally, we want to obtain imputed data matrix $\hat{X} \in \mathbb{R}^{n \times d}$ given $\tilde{X}$ and M where f is the imputation method:

$$\hat{X} = f(\tilde{X}, M) \tag{3}$$

4 Methodology

4.1 Overall Architecture

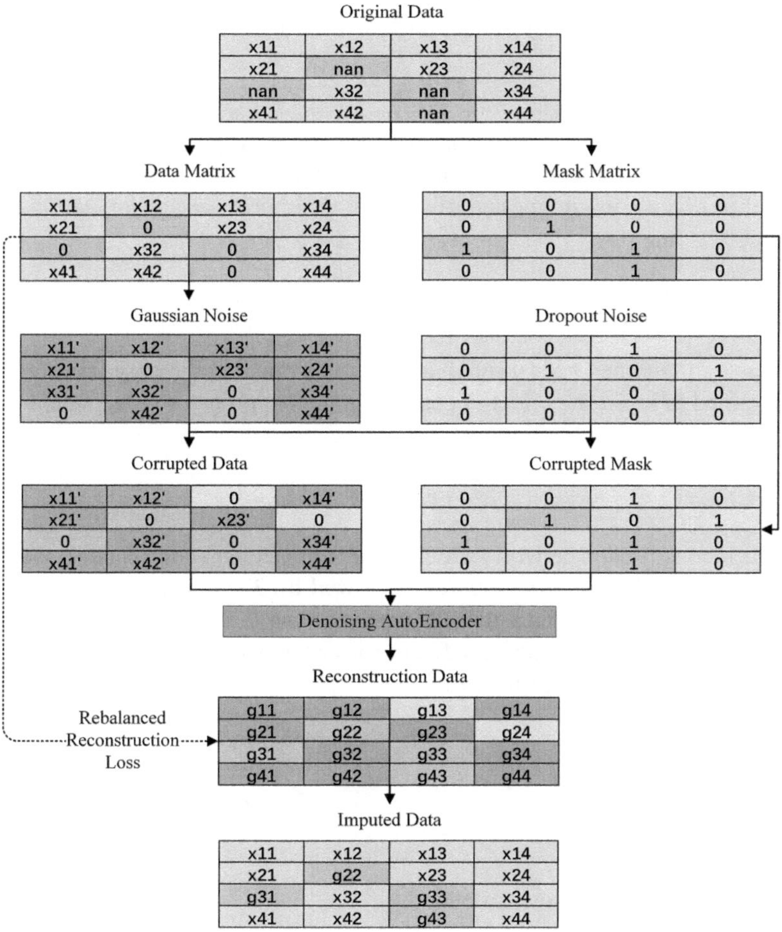

Fig. 1. DC-DAE architecture overview.

As shown in Fig. 1, the DC-DAE framework consists of two key phases: 1) concurrent corruption of the input data with masking and noise, and 2) reconstruction of the original complete data using a DAE architecture.

4.2 Masked Denoising

A key innovation of DC-DAE is jointly utilizing two different corruption techniques during training:

- Noise Injection: Adding Gaussian random noise ε to the observed values. The noise scale hyperparameter γ determines the noise level. $\varepsilon \sim \mathcal{N}(0, \gamma)$

$$\dot{X} = \tilde{X} + \varepsilon * (1 - M) \tag{4}$$

- Masking: Randomly dropout, or "masking", a subset of the observed feature values to create artificial missingness. The mask ratio hyperparameter β controls the amount of masking. Let $\overline{M}$ be the artificial mask matrix, then

$$\ddot{X} = \dot{X} * \left(1 - \overline{M}\right) \tag{5}$$

Here, $*$ operation denotes element-wise multiplication. Applying these corruptions prevents the model from directly accessing clean observed data. Additionally, it exposes the model to diverse missingness during training, improving generalization to unseen patterns at test time.

4.3 Reconstruction

The corrupted input then goes through a DAE to reconstruct the original complete data. DAE is trained to invert the effect of corruption by mapping corrupted samples to clean samples.

$$\overline{X} = \hat{f}\left(\ddot{X}, \overline{M} * M\right) \tag{6}$$

$\hat{f}$ is the reconstruction network, which architecture consists of 3 encoder layers which map the input to a latent representation, followed by 3 symmetric decoder layers that reconstruct the output from the code. All layers use tanh activation and the dimensionality of the hidden layers matches the input feature dimension. Combining with observed data, DC-DAE final imputation results are:

$$\hat{X} = \overline{X} * M + \tilde{X} * (1 - M) \tag{7}$$

4.4 Loss Function

The overall reconstruction loss is defined as:

$$L = \alpha * L_n + (1 - \alpha) * L_m \tag{8}$$

$$L_n = \left\| \left(\overline{X} - \tilde{X}\right) * \left(1 - \overline{M}\right) * (1 - M) \right\| \tag{9}$$

$$L_m = \left\| \left(\overline{X} - \tilde{X}\right) * \overline{M} * (1 - M) \right\| \tag{10}$$

Here, L_m penalizes reconstruction error on the artificially masked entries, while L_n penalizes error on noising observed values. The α parameter controls the balance between the two objectives. Setting $\alpha < 0.5$ puts less emphasis on reconstructing the noisy observed data. We demonstrate experimentally that this rebalancing improves imputation accuracy by preventing overfitting to the artificial masking distribution. The loss hyperparameters are tuned on a validation set.

5 Experiments

5.1 Datasets

We evaluate DC-DAE on 5 real-world tabular datasets from the UCI repository [35]. As is shown in Table 1, these datasets cover a diverse range of sizes and dimensions to comprehensively evaluate performance between DC-DAE and various state-of-the-art imputation methods. We remove the ID column in the Glass and Breast datasets and samples contain NA values in Breast dataset.

Table 1. Datasets used for evaluation

Dataset Full Name	Dataset Acronym	Samples	Features
Glass Identification	Glass	214	9
Breast Cancer Wisconsin (Original)	Breast	683	9
Spambase	Spam	4601	57
Letter Recognition	Letter	20000	16
Statlog (Shuttle)	Shuttle	58000	8

5.2 Evaluation Protocol

We introduce missingness artificially via missing completely at random (MCAR) masking. The datasets are split 70/30 into train/test. Missing values in the test set are imputed and evaluated. Normalization uses train set statistics.

Performance is measured by normalized root mean squared error (NRMSE) on the missing test values across 10 random missingness trials. We report average NRMSE over these trials. NRMSE is a common imputation evaluation metric used by baseline methods like DAEMA.

5.3 Implementation Details

The hyperparameters are set as:

- Mask ratio β: set to be 0.2 based on common usage in baseline methods.
- Noise scale γ: set to be 0.2 for simplicity.

- Denoising loss weight α: Tuned on validation data across the range 0.1 to 0.9. Optimal values found were 0.1 (Breast, Spam), 0.2 (Letter, Shuttle), 0.9 (Glass).

The model architecture uses 6 fully-connected layers with tanh activations and hidden dimensionality matching the input features. We select 6 layers to match the default depth required by the MIDA baseline method and to enable fair comparison by allowing other baselines to use the same capacity if desired. Models are trained with batch size 64 for 10000 epochs, using the Adam optimizer with a learning rate of 0.001.

5.4 Baseline

We compare DC-DAE with several state-of-the-art imputation methods:

- Traditional: Mean, MICE
- GAN-based: GAIN, IM-GAN
- DAE-based: MIDA, DAEMA
- VAE-based: HI-VAE, MIWAE

Recent works [36, 37] have identified challenges in reproducing and fairly comparing deep learning imputation algorithms due to inconsistencies in evaluation protocols and difference in model capacity. To mitigate this, we implement a standardized experiment procedure across all algorithms. The same train/test splits, missingness patterns, and evaluation metrics are applied to each method. Additionally, baseline architectures are adjusted to match the capacity of DC-DAE where possible, while retaining their core characteristics:

- For GAIN, we use equal hidden layer sizes rather than bottleneck layers and construct a 6-layer generator and a 3-layer discriminator.
- For IM-GAN, we replace BRNN cells with fully-connected networks in the same 6 + 3 architecture as GAIN.
- For DAEMA, we reduce the hidden size from [2d, 2d] to [d, 1] to prevent overparameterization and set the depth of feature decoder to 3.
- For MIWAE and HI-VAE, the decoder depth is set to 3 layers.
- For MIDA, we use the default hyperparameters from the original papers.
- For MICE, we utilize the IterativeImputer module from the scikit-learn Python package with its default settings.

This standardized protocol enables direct comparison between DC-DAE and state-of-the-art baselines.

6 Results

The experimental results show that the proposed DC-DAE method has achieved superior imputation performance compared to existing deep learning techniques. This validates our hypothesis that the dual corruption technique and balanced loss can enhance the generalization ability for various missing patterns. The results are consistent with the design objectives, with DC-DAE achieving a balance of accuracy and efficiency. It demonstrates the state-of-the-art performance of our proposed DC-DAE method against existing deep learning techniques for missing data imputation. Furthermore, the baseline comparisons in Table 2 highlight that DC-DAE achieves lower NRMSE than GAN-based, AE-based, and VAE-based approaches across diverse datasets and missingness ratios.

7 Analysis

To evaluate the contribution of each component of our algorithm, we conduct in-depth ablation studies among low, middle and high missingness on the Letter dataset. Furthermore, to investigate the flexibility of our plug-and-play approach, we incorporate various techniques such as cross-attention mechanism, adversarial training structure and overcomplete hidden space with DC-DAE.

7.1 Source of Gain

Table 3 shows that the imputation performance of DC-DAE is improved when dual corruptions and rebalanced loss are included. More specifically, the full DC-DAE has a 12% improvement over the vanilla AE (i.e. DC-DAE w/o α&L_n&L_m). Furthermore, tuning the loss rebalancing hyperparameter α gives improvement of 2% across different natural missing rates (i.e. DC-DAE w/o α), which controls the relative importance of reconstructing the artificially introduced missingness versus denoising the randomly noised observed data.

7.2 Cross-Attention, GAN and Overcomplete

As shown in Table 4, incorporating cross-attention between the data and mask provides no benefit, even decreasing performance in some missing rates (i.e. DC-DAE w/ attention). This indicates that the additional complexity is unneeded for DC-DAE's straightforward reconstructive approach. Similarly, introducing an adversarial discriminator offers gains only in middle missingness (i.e. DC-DAE w/ GAN). This aligns with the finding that flexible generative modelling is unnecessary for the primary task of missing value imputation. However, expanding the hidden layer sizes approximating MIDA and VAE architectures consistently improves results, especially for low missingness (i.e. DC-DAE w/ MIDA and DC-DAE w/ VAE). This highlights the value of increased representational capacity despite DC-DAE's lightweight architecture.

8 Future Works

While the proposed DC-DAE method achieves state-of-the-art imputation accuracy, several promising directions remain for future work:

- As the auxiliary denoising task provides benefits, exploring more advanced multi-task learning approaches like multi-gate mixture-of-experts [38] could further improve performance.

Table 2. Imputation performance (Mean ± Std of NRMSE) comparison between DC-DAE and baseline methods

	Glass	Breast	Spam	Letter	Shuttle
	Low miss rate: 0.2				
Mean	$1.003_{\pm 0.112}$	$0.993_{\pm 0.035}$	$1.007_{\pm 0.039}$	$0.997_{\pm 0.005}$	$0.969_{\pm 0.077}$
MICE	$0.757_{\pm 0.152}$	$0.679_{\pm 0.037}$	$1.051_{\pm 0.188}$	$0.746_{\pm 0.006}$	$\mathbf{0.573_{\pm 0.121}}$
GAIN	$1.221_{\pm 0.277}$	$1.298_{\pm 0.078}$	$1.196_{\pm 0.034}$	$1.297_{\pm 0.041}$	$0.928_{\pm 0.093}$
IM-GAN	$1.464_{\pm 0.330}$	$1.291_{\pm 0.694}$	$0.980_{\pm 0.040}$	$0.994_{\pm 0.031}$	$0.830_{\pm 0.088}$
MIDA	$0.798_{\pm 0.090}$	$0.736_{\pm 0.031}$	$0.942_{\pm 0.038}$	$0.802_{\pm 0.006}$	$0.802_{\pm 0.091}$
DAEMA	$\mathbf{0.739}_{\pm 0.093}$	$0.685_{\pm 0.050}$	$0.950_{\pm 0.040}$	$0.709_{\pm 0.010}$	$0.595_{\pm 0.117}$
HI-VAE	$0.963_{\pm 0.179}$	$0.766_{\pm 0.042}$	$1.027_{\pm 0.040}$	$0.750_{\pm 0.011}$	$0.773_{\pm 0.100}$
MIWAE	$0.903_{\pm 0.125}$	$0.759_{\pm 0.047}$	$1.014_{\pm 0.044}$	$0.742_{\pm 0.028}$	$0.674_{\pm 0.122}$
DC-DAE (ours)	$0.788_{\pm 0.118}$	$\mathbf{0.661_{\pm 0.041}}$	$\mathbf{0.927_{\pm 0.040}}$	$\mathbf{0.669_{\pm 0.006}}$	$0.595_{\pm 0.118}$
	Medium miss rate: 0.5				
Mean	$0.998_{\pm 0.089}$	$1.007_{\pm 0.020}$	$0.993_{\pm 0.029}$	$0.999_{\pm 0.003}$	$0.982_{\pm 0.040}$
MICE	$1.130_{\pm 0.168}$	$0.843_{\pm 0.070}$	$1.225_{\pm 0.200}$	$0.856_{\pm 0.009}$	$0.934_{\pm 0.135}$
GAIN	$1.534_{\pm 0.470}$	$1.285_{\pm 0.124}$	$1.113_{\pm 0.030}$	$1.378_{\pm 0.050}$	$1.256_{\pm 0.065}$
IM-GAN	$2.267_{\pm 0.605}$	$1.469_{\pm 0.309}$	$1.036_{\pm 0.027}$	$1.062_{\pm 0.025}$	$1.041_{\pm 0.122}$
MIDA	$0.958_{\pm 0.060}$	$0.863_{\pm 0.026}$	$\mathbf{0.966_{\pm 0.026}}$	$0.934_{\pm 0.003}$	$0.924_{\pm 0.043}$
DAEMA	$0.965_{\pm 0.079}$	$0.731_{\pm 0.033}$	$0.993_{\pm 0.026}$	$0.804_{\pm 0.005}$	$0.737_{\pm 0.052}$
HI-VAE	$0.943_{\pm 0.083}$	$0.824_{\pm 0.049}$	$1.015_{\pm 0.029}$	$0.853_{\pm 0.008}$	$0.857_{\pm 0.043}$
MIWAE	$\mathbf{0.940}_{\pm 0.093}$	$0.812_{\pm 0.036}$	$1.006_{\pm 0.031}$	$0.832_{\pm 0.026}$	$0.883_{\pm 0.047}$
DC-DAE (ours)	$0.942_{\pm 0.084}$	$\mathbf{0.717_{\pm 0.053}}$	$0.970_{\pm 0.036}$	$\mathbf{0.786_{\pm 0.006}}$	$\mathbf{0.730_{\pm 0.055}}$
	High miss rate: 0.8				
Mean	$1.035_{\pm 0.058}$	$1.006_{\pm 0.014}$	$1.000_{\pm 0.014}$	$1.000_{\pm 0.002}$	$0.991_{\pm 0.018}$
MICE	$1.280_{\pm 0.202}$	$1.010_{\pm 0.052}$	$1.185_{\pm 0.221}$	$0.983_{\pm 0.012}$	$1.280_{\pm 0.430}$
GAIN	$1.736_{\pm 0.578}$	$1.386_{\pm 0.057}$	$1.105_{\pm 0.015}$	$1.365_{\pm 0.033}$	$1.303_{\pm 0.022}$
IM-GAN	$3.233_{\pm 0.709}$	$2.349_{\pm 0.377}$	$1.499_{\pm 0.124}$	$2.264_{\pm 0.285}$	$2.325_{\pm 0.613}$
MIDA	$\mathbf{1.034}_{\pm 0.052}$	$0.977_{\pm 0.015}$	$\mathbf{0.995_{\pm 0.013}}$	$0.990_{\pm 0.002}$	$0.983_{\pm 0.017}$
DAEMA	$1.272_{\pm 0.189}$	$0.861_{\pm 0.023}$	$1.086_{\pm 0.035}$	$0.920_{\pm 0.003}$	$0.904_{\pm 0.019}$
HI-VAE	$\mathbf{1.034}_{\pm 0.034}$	$0.910_{\pm 0.035}$	$1.024_{\pm 0.015}$	$0.965_{\pm 0.025}$	$0.967_{\pm 0.029}$
MIWAE	$1.042_{\pm 0.034}$	$0.887_{\pm 0.030}$	$1.023_{\pm 0.016}$	$0.931_{\pm 0.002}$	$0.966_{\pm 0.024}$
DC-DAE (ours)	$1.191_{\pm 0.162}$	$\mathbf{0.839_{\pm 0.029}}$	$1.029_{\pm 0.015}$	$\mathbf{0.914_{\pm 0.003}}$	$\mathbf{0.893_{\pm 0.019}}$

Table 3. Source of gains in DC-DAE algorithm (Mean ± Std of NRMSE (GAIN (%)))

Miss Rate	0.2	0.5	0.8
DC-DAE (ours)	**0.669$_{\pm 0.006}$**	**0.786$_{\pm 0.006}$**	**0.914$_{\pm 0.003}$**
DC-DAE w/o α	0.703$_{\pm 0.009}$ (4.8%)	0.801$_{\pm 0.005}$ (1.9%)	0.919$_{\pm 0.003}$ (0.5%)
DC-DAE w/o α&L_n	0.707$_{\pm 0.008}$ (5.3%)	0.805$_{\pm 0.004}$ (2.3%)	0.921$_{\pm 0.004}$ (0.7%)
DC-DAE w/o α&L_m	0.808$_{\pm 0.011}$ (17.1%)	0.854$_{\pm 0.009}$ (7.9%)	0.951$_{\pm 0.007}$ (3.9%)
DC-DAE w/o α&L_n&L_m	0.837$_{\pm 0.025}$ (20.1%)	0.872$_{\pm 0.015}$ (9.9%)	0.962$_{\pm 0.012}$ (5.0%)

Table 4. Additional analysis of DC-DAE algorithm (Mean ± Std of NRMSE (GAIN (%)))

Miss Rate	0.2	0.5	0.8
DC-DAE (ours)	0.669$_{\pm 0.006}$	0.786$_{\pm 0.006}$	0.914$_{\pm 0.003}$
DC-DAE w/ attention	0.671$_{\pm 0.008}$ (-0.3%)	0.786$_{\pm 0.005}$ (0.0%)	0.916$_{\pm 0.005}$ (-0.2%)
DC-DAE w/ GAN	0.670$_{\pm 0.007}$ (-0.1%)	0.785$_{\pm 0.004}$ (0.1%)	0.915$_{\pm 0.003}$ (-0.1%)
DC-DAE w/ MIDA	**0.620**$_{\pm 0.006}$ (7.4%)	**0.754**$_{\pm 0.002}$ (4.0%)	**0.911**$_{\pm 0.004}$ (0.3%)
DC-DAE w/ VAE	0.658$_{\pm 0.006}$ (1.7%)	0.775$_{\pm 0.005}$ (1.4%)	0.912$_{\pm 0.003}$ (0.2%)

- Applying more sophisticated methods to handle mixed data types with both numerical and categorical features could enhance applicability to broader tabular dataset.
- Tuning the hyperparameters like mask ratio, nose scale and denoising loss weight simultaneously in a dataset-specific or automated manner could improve robustness across diverse missing data scenarios.
- Incorporating overcomplete representations and greater model capacity could potentially build upon the benefits of the masking denoising technique.

By addressing these limitations and extensions, the power of the simple yet effective masked denoising approach could be further leveraged to tackle missing data challenges.

9 Conclusions

This work introduces DC-DAE, a straightforward yet powerful technique for missing data imputation. DC-DAE corrupts input data during training through simultaneous masking and additive noise. We demonstrate experimentally that this prevents overfitting to any one artificial missingness pattern, facilitating imputation under real-world missingness conditions at test time.

The balanced loss function further enhances the approach by controlling the trade-off between reconstructing artificial missingness and denoising observed values. Our experiments across diverse datasets and missingness scenarios demonstrate state-of-the-art performance compared to existing deep learning imputation methods based on GANs, AEs, and VAEs.

The plug-and-play masked denoising technique provides an intuitive and effective solution for handling missing data. By combining simple corruption robustness with reconstructive modelling, DC-DAE achieves both accuracy and efficiency. The strong empirical results highlight the value of proper inductive bias when designing deep learning architectures for missing data problems.

Acknowledgment. This work is supported by GeoGPT Model and GeoCloud Platform of Zhejiang Lab (Project No. 3200-3AA240200).

References

1. Little, R.J., et al.: The prevention and treatment of missing data in clinical trials. N. Engl. J. Med. **367**(14), 1355–1360 (2012)
2. Yoon, J., Davtyan, C., van der Schaar, M.: Discovery and clinical decision support for personalized healthcare. IEEE J. Biomed. Health Inf. **21**(4), 1133–1145 (2016)
3. Duan, Y., et al.: A deep learning based approach for traffic data imputation. In: 17th International IEEE Conference on Intelligent Transportation Systems (ITSC). IEEE (2014)
4. Wooldridge, J.M.: Inverse probability weighted estimation for general missing data problems. J. Econometrics **141**(2), 1281–1301 (2007)
5. Lall, R.: How multiple imputation makes a difference. Polit. Anal. **24**(4), 414–433 (2016)
6. Jaseena, K.U., David, J.M.: Issues, challenges, and solutions: big data mining. CS IT-CSCP **4**(13), 131–140 (2014)
7. Camino, R.D., Hammerschmidt, C.A., State, R.: Improving missing data imputation with deep generative models. arXiv preprint arXiv:1902.10666 (2019)
8. Mehrabani-Zeinabad, K., Doostfatemeh, M., Ayatollahi, S.M.T.: An efficient and effective model to handle missing data in classification. Biomed. Res. Int. **2020**(1), 8810143 (2020)
9. Van Buuren, S., Groothuis-Oudshoorn, K.: mice: Multivariate imputation by chained equations in R. J. Stat. Softw. **45**, 1–67 (2011)
10. Raymond, M.R., Roberts, D.M.: A comparison of methods for treating incomplete data in selection research. Educ. Psychol. Measur. **47**(1), 13–26 (1987)
11. Strike, K., El Emam, K., Madhavji, N.: Software cost estimation with incomplete data. IEEE Trans. Softw. Eng. **27**(10), 890–908 (2001)
12. Acuna, E., Rodriguez, C.: The treatment of missing values and its effect on classifier accuracy. In: Classification, Clustering, and Data Mining Applications: Proceedings of the Meeting of the International Federation of Classification Societies (IFCS), Illinois Institute of Technology, Chicago, 15–18 July 2004. Berlin, Heidelberg: Springer Berlin Heidelberg (2004)
13. Yoon, J., Jordon, J., Schaar, M.: GAIN: missing data imputation using generative adversarial nets. In: International conference on machine learning. PMLR (2018)
14. Neves, D.T., et al.: From missing data imputation to data generation. J. Comput. Sci. **61**, 101640 (2022)
15. Gulrajani, I., et al.: Improved training of wasserstein GANs. Adv. Neural Inf. Process. Syst. **30** (2017)
16. Salimans, T., et al.: Improved techniques for training GANs. Adv. Neural Inf. Process. Syst. **29** (2016)
17. Vincent, P., et al.: Extracting and composing robust features with denoising autoencoders. In: Proceedings of the 25th International Conference on Machine Learning (2008)
18. Kingma, D.P., Welling, M.: Auto-encoding variational bayes. arXiv preprint arXiv:1312.6114 (2013)

19. Gondara, L., Wang, K.: MIDA: multiple imputation using denoising autoencoders. In: Advances in Knowledge Discovery and Data Mining: 22nd Pacific-Asia Conference, PAKDD 2018, Melbourne, VIC, Australia, June 3–6, 2018, Proceedings, Part III 22. Springer International Publishing (2018)
20. Goodfellow, I., et al.: Generative adversarial nets. In: Advances in Neural Information Processing Systems, vol. 27 (2014)
21. Zhang, W., et al.: Missing data repairs for traffic flow with self-attention generative adversarial imputation net. IEEE Trans. Intell. Transp. Syst. **23**(7), 7919–7930 (2021)
22. Li, S.C.-X., Jiang, B., Marlin, B.: MisGAN: Learning from incomplete data with generative adversarial networks. arXiv preprint arXiv:1902.09599 (2019)
23. Luo, Y., Zhang, Y., Cai, X., Yuan, X.: E2GAN: end-to-end generative adversarial network for multivariate time series imputation. In: Proceedings of the Twenty-Eighth International Joint Conference on Artificial Intelligence, IJCAI-19, Macao, China, 10–16 August, pp. 3094–3100 (2019)
24. Hwang, U., Jung, D., Yoon, S.: HexaGAN: generative adversarial nets for real world classification. International Conference on Machine Learning. PMLR, (2019)
25. Yang, B., et al.: ST-LBAGAN: spatio-temporal learnable bidirectional attention generative adversarial networks for missing traffic data imputation. Knowl. Based Syst. **215**, 106705 (2021)
26. Yoon, S., Sull, S.: GAMIN: generative adversarial multiple imputation network for highly missing data. In: Proceedings of the IEEE/CVF Conference on Computer Vision and Pattern Recognition (2020)
27. Costa, A.F., et al.: Missing data imputation via denoising autoencoders: the untold story. In: Advances in Intelligent Data Analysis XVII: 17th International Symposium, IDA 2018,'s-Hertogenbosch, The Netherlands, October 24–26, 2018, Proceedings 17. Springer International Publishing (2018)
28. Sánchez-Morales, A., et al.: Improving deep learning performance with missing values via deletion and compensation. Neural Comput. Appl. **32**, 13233–13244 (2020)
29. Hallaji, E., Razavi-Far, R., Saif, M.: DLIN: deep ladder imputation network. IEEE Trans. Cybern. **52**(9), 8629–8641 (2021)
30. Nazabal, A., et al.: Handling incomplete heterogeneous data using vaes. Pattern Recogn. **107**, 107501 (2020)
31. Mattei, P.-A., Frellsen, J.: MIWAE: deep generative modelling and imputation of incomplete data sets. International Conference on Machine Learning. PMLR (2019)
32. Wu, R., et al.: Attention-based learning for missing data imputation in HoloClean. Proc. Mach. Learn. Syst. **2**, 307–325 (2020)
33. Tihon, S., et al.: DAEMA: denoising autoencoder with mask attention. In: International Conference on Artificial Neural Networks. Cham: Springer International Publishing (2021)
34. Wu, Z., et al.: Imputing missing indoor air quality data with inverse mapping generative adversarial network. Build. Environ. **215**, 108896 (2022)
35. Asuncion, A., Newman, D.: UCI machine learning repository (2007)
36. Camino, R.D., Hammerschmidt, C.: Working with deep generative models and tabular data imputation. First Workshop on the Art of Learning with Missing Values (Artemiss) (2020)
37. Lin, W.-C., Tsai, C.-F.: Missing value imputation: a review and analysis of the literature (2006–2017). Artif. Intell. Rev. **53**, 1487–1509 (2020)
38. Ma, J., et al.: Modeling task relationships in multi-task learning with multi-gate mixture-of-experts. In: Proceedings of the 24th ACM SIGKDD International Conference on Knowledge Discovery & Data Mining (2018)

Remote Intelligent Inspection Method of Man-Machine Interactive Computer Room Based on Digital Twin Technology

Xuyuan Kang(✉), Lin Long, Ran Li, Linlin Zhang, Meiqing Yang, and Qiang Tong

Three Gorges Cascade Dispatch & Communication Center of China Yangtze Power Co., Ltd., YiChang 443000, HuBei, China
kang_xuyuan@ctg.com.cn

Abstract. In order to ensure the normal operation of the equipment in the man-machine interactive computer room, a remote intelligent inspection method of the man-machine interactive computer room based on digital twinning technology is proposed. Collect information data in that remote intelligent inspection process of the man-machine interactive machine room, standardize and unifying the information in the environmental inspection process, fitting the characteristic data of the signal information, the equipment information and the environmental information of the man-machine interactive machine room, and obtain the collection result of the remote intelligent inspection information data of the man-machine interactive machine room. According to the spatial distribution value, the change of standard temperature value is calculated, and the digital twin technology inspection model is constructed to realize the remote intelligent inspection of man-machine interactive computer room. The experimental results show that this method has good inspection effect, can ensure the accurate tracking of driving direction and angle, and has high inspection accuracy.

Keywords: Digital twinning technology · Man-machine interaction · Computer room · Remote intelligent inspection

1 Introduction

With the rapid development of information technology, data centers and computer rooms have become an important part of information infrastructure in modern society [1]. The stable operation of equipment in the computer room is directly related to the efficiency of enterprise operation and the security of data. However, the traditional inspection method of computer room often relies on manual inspection, which is not only inefficient, but also difficult to achieve comprehensive and accurate monitoring in the face of complex equipment and massive data [2, 3].

If it is determined that the target needs to be tracked in real time, the initial task allocation and parameter adjustment of the load are carried out, and the task target is tracked continuously to realize the rapid response of the edge crisis. If there is a flight

G. Jeon and X. Kong (Eds.): ACIIBD 2024, LNICST 616, pp. 36–42, 2025.
https://doi.org/10.1007/978-3-031-91853-7_4

route adjustment instruction, the adjustment route will be sent to the flight controller, which greatly improves the flexibility and efficiency of dealing with the crisis during the inspection of the drone. By adopting this early warning method, users can be provided with newer, faster and more efficient ways to realize the early warning demand in UAV inspection, which is a new breakthrough in the field of UAV command methods. On the basis of the above research, this paper puts forward a remote intelligent inspection method of computer room based on digital twinning technology.

2 Remote Intelligent Inspection of Man-Machine Interactive Computer Room Under Digital Twinning Technology

2.1 Man-Machine Interactive Computer Room Remote Intelligent Inspection Information Data Collection

Fitting the characteristic data of the signal information, equipment information and environmental information of the man-machine interactive computer room, thus obtaining the collection result of the remote intelligent inspection information data of the man-machine interactive computer room [4, 5], and the formula is as follows:

$$R_G = \lambda_1 Y_y + \lambda_2 \mu_n + \lambda_3 S_q \tag{1}$$

In the formula, λ_1, λ_2 and λ_3 respectively represent the weights corresponding to equipment information, material information and environmental information, and $\lambda_1 + \lambda_2 + \lambda_3 = 1$.

Formula (1) can be used to complete the remote intelligent inspection information data collection of man-machine interactive computer room, which lays a solid foundation for the subsequent construction of digital twin technology inspection model.

2.2 Digital Twin Technology Inspection Model Establishment

Construct a structural matrix of the environmental parameters of the man-machine interactive computer room, and record it as $W = \{w_1, w_2, ..., w_n\}$ [6, 7], indicating the number of environmental data, and get the spatial distribution values of different environmental parameters in the physical field, namely:

$$W = \begin{bmatrix} W_1^1 & W_1^2 & W_1^L \\ W_2^1 & W_2^2 & W_2^L \\ W_N^1 & W_N^2 & W_N^L \end{bmatrix} \tag{2}$$

In the formula, N represents the number of grid nodes; L represents the number of snapshots.

2.3 Realization of Remote Intelligent Inspection of Man-Machine Interactive Computer Room

In the process of remote intelligent inspection of man-machine interactive computer room, it is necessary to accurately control the inspection direction, based on accurate positioning data, which is obtained by Global Positioning System mobile station, and combined with navigation control to complete the inspection, so as to ensure that the inspection route of man-machine interactive computer room is the set route.

Suppose (x, y) represents the current position of the robot, θ represents the driving angle, $D_1(x_1, y_1)$ and $D_2(x_2, y_2)$ represent the starting point and the ending point respectively, and they are known routes and are on the center line. Δd and $\Delta \omega$ represent position deviation and angle deviation respectively.

In the formula, R_j represents the data of human-computer interactive computer room before normalization; R_j' represents the normalized human-computer interactive computer room data [8–10]; $\min(R_j)$ represents the minimum value of data in the human-computer interactive computer room; $\max(R_j)$ stands for the maximum value of data in human-computer interactive computer room.

Measure the similarity between human-computer interactive computer room data and abnormal data set, and the expression is:

$$\zeta(S_1, S_2) = \frac{|S_1 \cap S_2|}{|S_1 \cup S_2|} \tag{3}$$

In the formula, $\zeta(S_1, S_2)$ represents the similarity value between the data set S_1 and the abnormal data set S_2 of the human-computer interactive computer room. When the calculation result $\zeta(S_1, S_2)$ of formula (13) is less than 0.35, it is determined that the automatic man-machine interactive computer room is in normal operation; When the calculation result $\zeta(S_1, S_2)$ of formula (13) is greater than or equal to 0.35, it is determined that the operation state of the man-machine interactive computer room is abnormal, and corresponding alarm information is generated to remind the staff to check and deal with the panoramic view of the man-machine interactive computer room. The above process completes the remote intelligent inspection of man-machine interactive computer room, which provides more effective guarantee for safe operation.

3 Experimental Analysis

In order to verify the validity of the remote intelligent inspection method of man-machine interactive computer room based on digital twinning technology, simulation experiments are carried out, and the remote intelligent inspection method of man-machine interactive computer room based on digital twinning technology (the proposed method), the operation and maintenance inspection technology method of communication computer room based on AR (the method of reference [6]) and the intelligent early warning method based on UAV inspection (the method of reference [7]) are used to carry out the tests respectively. The resolution of visible light camera can be up to 1920×1080, and the frame rate is 6 ~ 35f/s; The resolution of infrared camera can be up to 320×240, and the frame rate is 1 ~ 20f/s; The frequency of WLAN is 2.5 Hz, the protocol is 802.11n, the bandwidth is 10 Mb/s, and the coverage diameter is not less than 200 m. In the above environment, configure the experimental parameters, as shown in Table 1.

Table 1. Table of Equipment Operation Parameters in Man-machine Interactive Computer Room

SERIAL NUMBER	INDEX	DATA
1	MAN-MACHINE INTERACTIVE ROOM PRESSURIZATION	> 10PA
2	RATED FREQUENCY	50HZ
3	POWER FACTOR	0.095
4	ELECTRICAL PROPERTY	2.595AC/M06Ω
5	ROLLING LOAD	136KG
6	DRAIN CURRENT	< 1MA
7	POWER SUPPLY CAPACITY	12-24V
8	CONTROLLER LINE LENGTH	30M

In the patrol operation, the accuracy and consistency of the patrol trajectory is very important to ensure the comprehensive coverage of the patrol area. In order to evaluate the actual effect of this method, the trajectory tracking performance of this method is tested in two cases: straight driving and angle turning driving. The test results are shown in Fig. 1. In Fig. 1(a), it is shown that the driving trajectory of a straight path is tracked and completed by the method in this paper at an inclination angle of 45. In Fig. 1(b), centering on the O point, it is verified that the method in this paper accurately tracks the inspection trajectory according to the preset angle.

From the analysis of Fig. 1, it can be seen that, under the condition of fixed included angle, the straight-line driving trajectory is highly consistent with the set trajectory, which means that the inspection effect of this method is good and the accuracy of the straight-line driving trajectory is guaranteed; With O point as the center point, the driving trajectory is in good agreement with the set trajectory within the range of 90, which fully shows that the method in this paper has good inspection effect, which can ensure the accurate tracking of driving direction and angle, and can ensure the tracking of inspection trajectory under different routes.

In order to further clearly show the application effect of the proposed method, the similarity $\zeta(S_1, S_2)$ between human-computer interactive computer room data and abnormal data set is selected as the evaluation index, and the $\zeta(S_1, S_2)$ value obtained through experiments is shown in Table 2.

According to Table 2, under different experimental working conditions, the similarity values of the data and abnormal data sets of man-machine interactive computer room obtained by this method are consistent with the actual results, while the similarity values of the methods of reference [6] and reference [7] are quite different from the actual results, which proves that the remote intelligent inspection accuracy of man-machine interactive computer room is higher by this method.

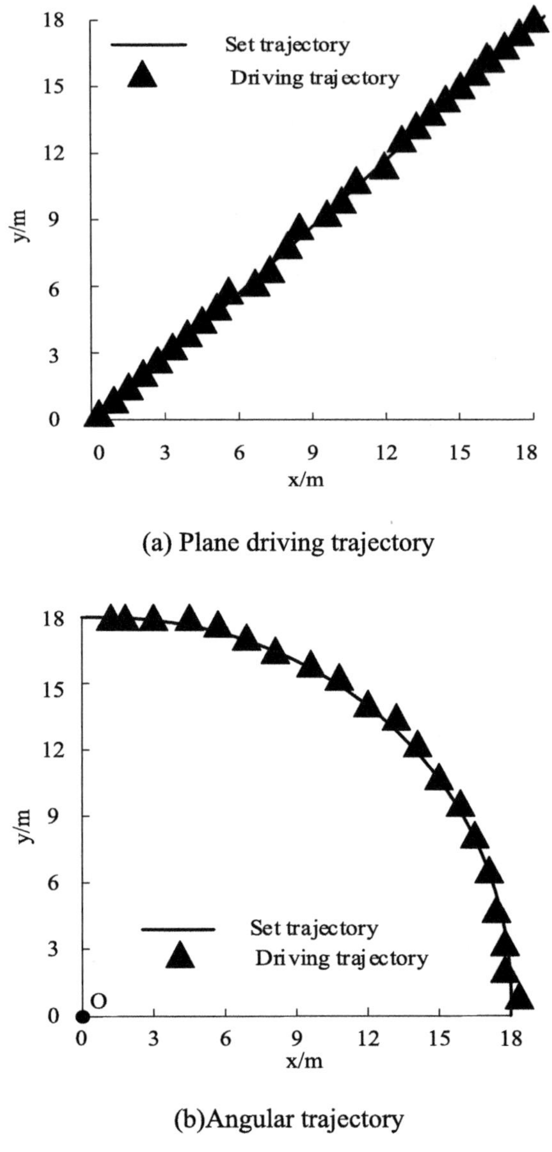

Fig. 1. Test results of inspection effect under this method.

Table 2. Numerical comparison results of similarity between human-computer interactive computer room data and abnormal data sets

WORKING CONDITION NUMBER	NET EFFECT	THE METHOD IN THIS PAPER	METHOD IN REFERENCE [6]	METHOD IN REFERENCE [7]
1	0.42	0.42	0.56	0.59
2	0.46	0.46	0.60	0.60
3	0.52	0.52	0.61	0.61
4	0.56	0.56	0.65	0.66
5	0.62	0.62	0.68	0.69
6	0.66	0.66	0.70	0.71
7	0.72	0.72	0.72	0.72
8	0.76	0.76	0.76	0.76
9	0.82	0.82	0.79	0.79
10	0.86	0.86	0.82	0.89

4 Conclusion

As one of the core infrastructures of enterprise information construction, the stability and security of data center computer room are particularly important. The traditional inspection method of computer room relies on manual inspection, which has some problems such as low efficiency, slow response speed and great security risks. Therefore, this paper puts forward a remote intelligent inspection method of computer room based on digital twinning technology, which is of great significance to improve the management level and operation and maintenance efficiency of computer room. Through experiments, we get the following results:

(1) The method in this paper has a good inspection effect, which can ensure the accurate tracking of driving direction and angle.
(2) The method in this paper has high intelligent inspection accuracy.

Looking forward to the future, it will continue to develop in the following aspects:

(1) With the continuous development of artificial intelligence technology, the intelligent inspection system will further realize intelligence and autonomy. By introducing more advanced algorithms and models, the accurate prediction and early warning of the equipment in the computer room can be realized, and the efficiency and safety of the operation and maintenance of the computer room can be improved.
(2) In computer room management, there are many types and sources of data involved. Future research will further strengthen the integration and processing ability of multi-source data, and realize comprehensive perception and accurate control of computer room environment.

(3) At present, the intelligent inspection method based on digital twin technology is mainly applied to the operation and maintenance management of data center computer rooms. In the future, the research of this method can be extended to other fields, such as smart city and intelligent manufacturing, and provide new solutions for operation and maintenance management in different fields.

Acknowledgements. The project was funded by the China Yangtze Power Co., Ltd. (Grant number: Z542302003).

References

1. Chernysheva, K.V., Azarkhin, A.V.: Features of the inspection of electronic media in the investigation of fraud in the field of computer information. Aktual'nye Probl. Pravovedeniya **3**(75), 32–35 (2022)
2. Edwards, C., Vaske, A., Mcdaniel, N., et al.: Real-time change detection for automated test socket inspection using advanced computer vision and machine learning. IEEE Trans. Semicond. Manufact. **36**(3), 332–339 (2023). A Publication of the IEEE Components, Hybrids, and Manufacturing Technology Society, the IEEE Electron Devices Society, the IEEE Reliability Society, the IEEE Solid-State Circuits Council
3. Guan, N., Zhang, P., Suthakorn, J.: Active compliance control system for intelligent inspection robot in power room. J. Control Sci. Eng. **2022**(1), 1–7 (2022)
4. Mohammadi, M., Rashidi, M., Yu, Y.: (103881) Integration of TLS-derived Bridge Information Modeling (Br IM) with a Decision Support System (DSS) for digital twinning and asset management of bridge infrastructures. Comput. Ind. **147**(1), 1–13 (2023)
5. Gao, Z., Paul, A., Wang, X.: Guest editorial: digital twinning: integrating AI-ML and big data analytics for virtual representation. IEEE Trans. Industr. Inf. **18**(2), 1355–1358 (2022)
6. Zengyang, M., Mi, X., Dayang, W., et al.: Research on AR-based operation and maintenance patrol inspection technology of communication equipment room. Telecom Power Technol. **40**(4), 232–234 (2023)
7. Li, Q., Xu, Y.: Intelligent early warning method based on drone inspection. J. Uncertain Syst. **14**(4), 1–10 (2022)
8. Yadav, K.S., Kirupakaran, A.M., Laskar, R.H.: End-to-end bare-hand localization system for human-computer interaction: a comprehensive analysis and viable solution. Vis. Comput. **40**(2), 1145–1165 (2024)
9. Karahasanovi, A., Alma Leora, C.: Project-based learning in human–computer interaction: a service-dominant logic approach Int. Technol. Smart Educ. **18**(1), 1–10 (2023)
10. Zhao, H., Han, J.L., Hu, P., et al.: A digital twin-based method for equipment state inference in active power grids. Comput. Simul. **41**(2), 146–150 (2024)

Analysis of College Students' Achievement Based on Data Mining

Yimin Zhang(✉), Xiaolan Wu, Hongzhao Lv, Jieshan Sun, Jiarong Miao, and Ruofan Huang

Fujian Electric Power Vocational and Technical College, Quanzhou 362000, Fujian, China
18150298893@163.com

Abstract. Data mining, as an advanced data processing technology in the field of information science, has the core function of discovering valuable information from massive data sets. This process can not only reveal the intrinsic connection between data but also provide scientific and accurate guidance for decision-makers. This paper proposes a data mining method based on the C4.5 algorithm to mine the association rules of students' academic performance in a college so as to reveal the correlation between the performance and variables such as students' backgrounds and study habits, as well as the interdependencies between various courses. This analysis serves to inform the development of specialized talent cultivation programs and provides a scientific basis for the more strategic planning and execution of educational administration tasks.

Keywords: data mining · association rule · achievement analysis

1 Introduction

Data mining, synonymous with knowledge discovery, is defined as the process of extracting knowledge of potential value and utility to users from massive data sets. It plays a crucial role in transforming information into value by applying advanced algorithms and statistical methods to extract valuable patterns and insights from large amounts of data. Presently, data mining is predominantly utilized in fields such as healthcare, business intelligence, educational analytics, manufacturing engineering, and bioinformatics [1–6]. The implementation of data mining technology is typically integrated with various information technologies, leveraging highly automated data analysis processes and mechanisms for uncovering implicit rules, thereby assisting users in optimizing strategies and mitigating risks. R. Agrawal et al. introduced a classic algorithm for association rules, known as the Apriori algorithm [7], which is used to discover the relationship between frequent itemsets and association rules in a data set. The Apriori algorithm is characterized by its simplicity and ease of implementation, enabling it to manage relatively large datasets. However, due to its requirement to generate a substantial number of candidate item sets, the efficiency of the algorithm can be significantly diminished when dealing with massive datasets and high-dimensional data. Building upon this groundwork, H.

G. Jeon and X. Kong (Eds.): ACIIBD 2024, LNICST 616, pp. 43–50, 2025.
https://doi.org/10.1007/978-3-031-91853-7_5

Hamsa et al. have conducted an analytical and predictive study on student performance using a hybrid approach involving decision trees and fuzzy genetic algorithms [8]. However, the above methods have issues such as requiring training data cleaning, difficulty in processing high-dimensional data, and inconvenient understanding.

Therefore, this work proposes an analytical approach to student performance using the C4.5 decision tree algorithm [9, 10], which can readily handle high-dimensional data and is applicable to various types of datasets as well as being easy to understand and interpret.

2 Methodology

Data association [11] is an analytical method that explores the implicit interconnections among variables within a data set. The purpose of association analysis is to mine the complex web of relationships hidden between data to discover meaningful patterns and rules. Assume that $M = \{i_1, i_2, i_3, \ldots, i_m\}$ is the set of items, and the task-associated data K is the set of database transactions, where each transaction N is a set of items such that $N \subseteq M$. Each transaction is identified by a unique identifier, denoted as NMK. If A is an itemset, it can be expressed as $A \subseteq N$.

An association rule is an implication of the form $A \Rightarrow B$, where $A \subseteq M$, $B \subseteq M$, and $A \cap B = \emptyset$. Rule $A \Rightarrow B$ holds in transaction set K with a support of q, which is the percentage of transactions in K that contain the set $A \cup B$, and can be represented by the probability $P(A \cup B)$. Rule $A \Rightarrow B$ has a confidence d in the transaction set K, which d is the percentage of transactions in K that contain B given that they contain A, which can be expressed by the conditional probability $P(A|B)$. So

$$Support(A \Rightarrow B) = P(A \cup B) \quad (1)$$

$$Confidence(A \Rightarrow B) = P(A|B) \quad (2)$$

Rules that satisfy both the minimum support threshold (min_sup) and the minimum confidence threshold (min_conf) are called strong rules. For a given transactional database, association rule mining is the process of finding suitable association rules by using user-specified min_sup and min_conf.

Generally, the association rule mining problem can be divided into two aspects. The first aspect involves identifying all frequent itemsets that satisfy the min_sup, i.e., the frequency of occurrence of these itemsets at least meets a given support criterion. In practice, there may be containment relationships between frequent itemsets, but usually we are more interested in independent frequent large itemsets that are not contained by other frequent itemsets. These frequent large itemsets are the basis for further analysis as they provide the most significant patterns in the data set. The second aspect involves generating rules with strong associations from the identified frequent itemsets. These rules reveal meaningful connections between items in the data, typically assessed by metrics such as confidence and lift, which measure the strength of the rules.

In this study, the C4.5 algorithm, which is an evolution of the ID3 algorithm, diverges primarily in its criteria for attribute selection. The C4.5 algorithm uses the information gain ratio as a measure of attribute importance, whereas the ID3 algorithm is based on information gain. The information gain ratio accounts for the ratio of the information gain due to an attribute split to the complexity introduced by the attribute's own values, thereby enhancing the algorithm's accuracy and generalization capability. Therefore, the C4.5 algorithm can be considered as an improved version of the ID3 algorithm, which is more effective in dealing with datasets with continuous attributes and missing values. By default, the selection of the sample training dataset in data mining is subject to value replacement. a) For interval variables with missing values, the missing values are replaced by the mean of the corresponding variable's samples. b) Binary variables with missing values are transformed into nominal or ordinal variables. In the C4.5 algorithm, the selection of the splitting attribute is contingent upon the computation of the information gain ratio for each candidate attribute. The information gain ratio is the quotient of the information gain and the split information associated with the attribute, serving as an evaluation of the attribute's contribution to the enhancement of the dataset's purity. The information gain ratio can be written as

$$GainRatio(S, W) = \frac{Gain(S, W)}{IV(W)} \tag{3}$$

which

$$Gain(S, W) = Entropy(S) - \sum\nolimits_{v \in Value(W)} \frac{|S_v|}{|S|} \cdot Entropy(S_v) \tag{4}$$

$$IV(W) = -\sum\nolimits_{v \in Value(W)} \frac{|S_v|}{|S|} \log_2 \frac{|S_v|}{|S|} \tag{5}$$

where, $Gain(S, W)$ is the information gain, $IV(W)$ represents the intrinsic value of attribute W, and $Entropy(S)$ denotes the information entropy.

The C4.5 algorithmic procedure is delineated as follows:

1) Initiate the root node.
2) Should the class of a correspond to the class in the original dataset, they are classified as the same category.
3) In the absence of elements in the original list of feature attributes, attribute a with the most prevalent category from the original dataset.
4) Select an attribute and partition the original data into different subsets.
5) Regenerate a root node; if the resulting subsets are void, tag the newly generated node with the most frequent category from the original dataset.
6) If the subsets are not void, recursively execute the algorithmic steps.
7) Elevate the newly generated node to a child node under the root node.

The C4.5 algorithm improves on the ID3 algorithm in its own way. As far as the algorithm itself is concerned, the accuracy of C4.5 is relatively high, but due to the time it takes to execute multiple commands on the existing dataset, the algorithm is inefficient, and due to its computerized system, there are some errors in the selection of sample attributes. The C4.5 algorithm, when selecting the splitting attribute, calculates the information gain ratio for each candidate attribute and selects the attribute with the highest ratio as the splitting attribute for the node. If the information gain ratio falls below a preset threshold or if no attribute's information gain ratio shows a significant improvement, the algorithm ceases to split further, thereby avoiding overfitting and enhancing the model's generalization capability.

3 Results and Discussions

In this section, we mainly use python to complete the preparation of the programme, which is mainly applied to the data analysis tools are PyCharm.

Firstly, the data undergoes preprocessing. 1) Data integration is performed, which is the process of consolidating data. It encompasses four main components: basic personal information, course information, ordinary grades, and final grades, aiming to facilitate the subsequent data mining and analysis, as illustrated in Fig. 1. 2) The data employed in the study undergoes discretization, which involves classifying students' achievements based on their examination scores. Data discretization, also known as data binning, primarily includes three methods: equal-width binning, equal-frequency binning, and custom binning. In this research, equal-width binning is utilized to categorize student grades [12, 13]. The specific categorization results are presented in Table 1.

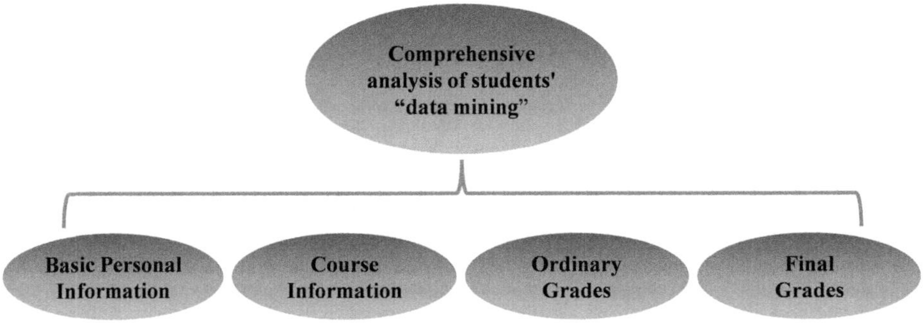

Fig. 1. Data Integration Operations.

Table 1. Grade Discretization Processing

Grade	Level	Source
A	Well	[90, 100]
B	Good	[80, 90]
C	Mid	[70, 80]
D	Pass	[60, 70]
E	Flunk	[0, 60]

Subsequently, our focus shifts to the extraction of characteristic factors and their impact on academic performance. Following a comprehensive processing of the dataset, an in-depth analysis of students' grades is imperative to pinpoint the key determinants that genuinely affect learning outcomes. Combining layered algorithms [14, 15], our study takes into account four extrinsic factors that may influence learning effectiveness: the students' age, the number of subject competitions they participated in, the degree of interaction in class, and the amount of time they spent studying after class. The predictive analysis results of these factors are clearly presented in Fig. 2. Through meticulous analysis and refined processing of the data, we are able to visually observe how each factor uniquely and variably affects students' academic achievements.

Examination of Fig. 2 reveals several insights into the factors influencing students' pass rates. In Fig. 2(a), students aged 19 to 23 consistently achieve a pass rate of over 88%, with relative stability in pass rates across different age groups. This suggests that age does not significantly affect the pass rates of students. Figure 2(b) presents the impact of the frequency of participation in academic competitions on pass rates. The bar chart shows minimal variance in heights across different frequencies, indicating that the number of academic competitions does not play a critical role in determining students' pass rates. The relationship between the degree of classroom interaction and pass rates is shown in Fig. 2. It can be deduced that a lower level of interaction is associated with a decreased pass rate. Passive learning in the classroom may reduce the pass rate to as low as 70%. Furthermore, Fig. 2(d) demonstrates the association between the utilization of after-class study time and pass rates. As the utilization of after-class time diminishes, so does the pass rate, highlighting the essential role of this factor in students' academic success. From the above, the degree of student interaction during class and the utilization of after-class study time can be considered as key feature factors to further monitor and assess the actual learning conditions of students in the classroom.

Finally, the inter-correlation prediction among courses is conducted. By applying the C4.5 algorithm to construct a model of inter-correlations between course grades, and training with the preprocessed dataset. We evaluate four distinct course models, denoted as Course_a, Course_b, Course_c, and Course_d, to analyze the degree of association among them, as illustrated in Table 2.

Table 2 reveals that the degree of association between Course_b and Course_d is the highest at 97%, while the association between Course_a and Course_d is comparatively lower, standing at only 84%. Utilizing these inter-course associations, we can predict students' course performance. In addition, by conducting a data mining analysis of a vast array of grades from students of various majors, we have identified courses with strong and meaningful correlations from a wide range of offerings. In conjunction with practical teaching requirements and societal demands, we have organized students' elective courses into modular structures and categorized them for elective purposes.

Following the model assessment, the effectiveness of the model has been established. To concretely verify the model's predictive accuracy, we manually input the actual and predicted grades for several students across two courses for comparison by means of the sampling algorithm [16]. As depicted in Fig. 3, we have entered the predicted grades for 10 students (red/blue solid line), arranged from lowest to highest. These are compared against their actual grades (red/blue dotted line). When the sample size of students is sufficiently large, the dashed line in the figure will nearly coincide with the solid line, indicating that the students' actual grades will approximate the predicted grades.

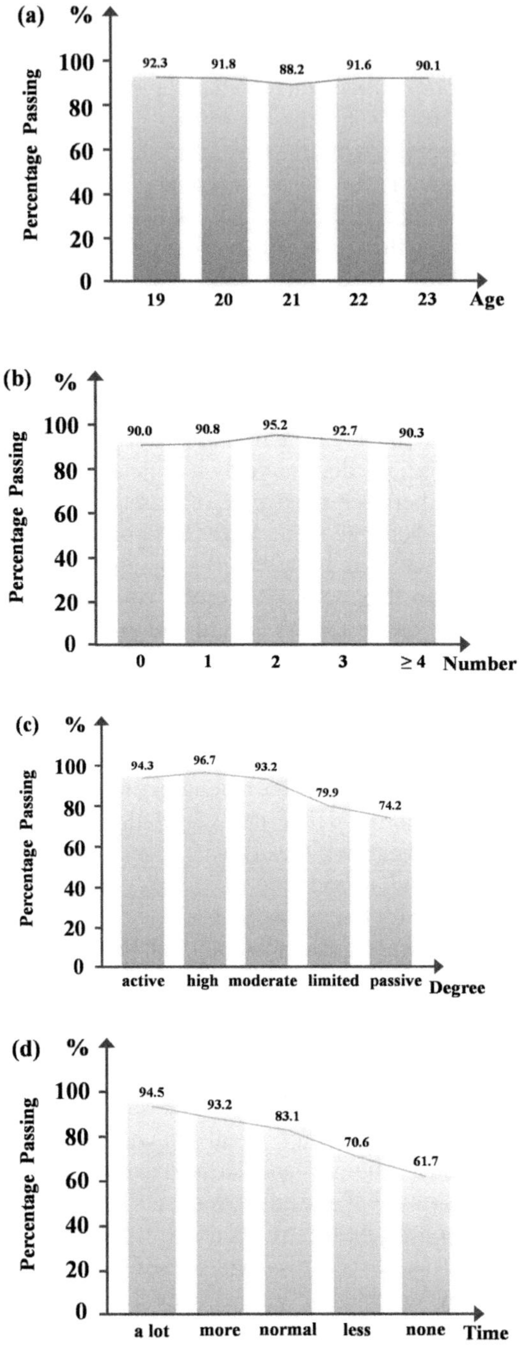

Fig. 2. The impact of different external factors on student's achievement.

Table 2. Prediction of Correlation Between Courses

Course	a	b	c	d
a		87%	96%	84%
b	87%		91%	97%
c	96%	91%		88%
d	84%	97%	88%	

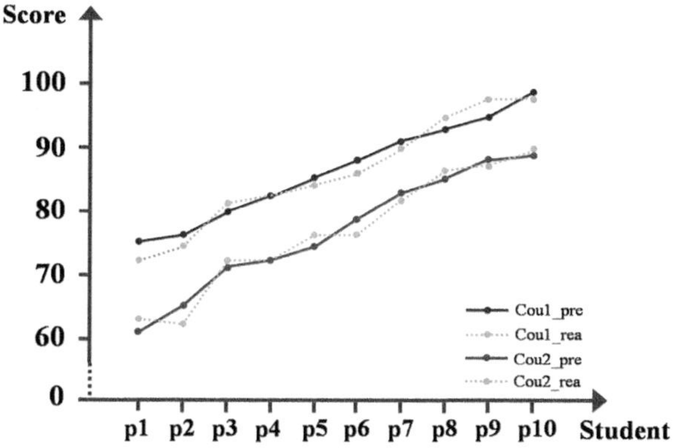

Fig. 3. Comparison of model predicted and actual grades plotted for 10 students' grades in each of two different courses.

4 Conclusions

In conclusion, we explored the most significant factors affecting students' achievement, and selected appropriate algorithms to build a prediction model conditional on these characteristic factors, so as to complete the prediction and analysis of students' performance. The student grade predictions and experimental comparisons were conducted in accordance with the mined course association rules. The comparative analysis between the predicted outcomes and actual data indicates that the algorithm and the experimental system possess significant instructional value, providing support for educational administrators.

Acknowledgements. This work is supported by the Fujian Electric Power Vocational and Technical College 2024 Academic Research Project: Application of Data Mining in Student Academic Performance (No. 2024KY010).

References

1. Alam, A.: Improving learning outcomes through predictive analytics: enhancing teaching and learning with educational data mining. In: 2023 7th International Conference on Intelligent Computing and Control Systems (ICICCS). IEEE, pp. 249–257 (2023)

2. Baek, C., Doleck, T.: Educational data mining versus learning analytics: a review of publications from 2015 to 2019. Interact. Learn. Environ. **31**(6), 3828–3850 (2023)
3. Adaga, E.M., Egieya, Z.E., Ewuga, S.K., Abdul, A.A., Abrahams, T.O.: Philosophy in business analytics: a review of sustainable and ethical approaches. Int. J. Manag. Entrepreneurship Res. **6**(1), 69–86 (2024)
4. Guo, Z., et al.: Diffusion models in bioinformatics and computational biology. Nat. Rev. Bioeng. **2**(2), 136–154 (2024)
5. Zhang, Y., Ghandour, A., Shestak, V.: Using learning analytics to predict students performance in moodle LMS (2020)
6. Dyckhoff, A.L., Zielke, D., Bültmann, M., Chatti, M.A., Schroeder, U.: Design and implementation of a learning analytics toolkit for teachers. J. Educ. Technol. Soc. **15**(3), 58–76 (2012)
7. Agrawal, R., Srikant, R.: Fast algorithms for mining association rules. In: Proceedings 20th International Conference Very Large Data Bases, VLDB, vol. 1215, pp. 487–499 (1994)
8. Hamsa, H., Indiradevi, S., Kizhakkethottam, J.J.: Student academic performance prediction model using decision tree and fuzzy genetic algorithm. Procedia Technol. **25**, 326–332 (2016)
9. Ruggieri, S.: Efficient C4. 5 [classification algorithm]. IEEE Trans. Knowl. Data Eng. **14**(2), 438–444 (2002)
10. Aldino, A.A., Sulistiani, H.: Decision tree c4. 5 algorithm for tuition aid grant program classification (case study: department of information system, Universitas Teknokrat Indonesia). Jurnal Ilmiah Edutic: Pendidikan dan Informatika **7**(1), 40–50 (2020)
11. Bar-Shalom, Y., Osborne, R.W.: Data association. Encyclopedia of Systems and Control. Cham: Springer International Publishing, pp. 509–517 (2021)
12. Savasere, A., Omiecinski, E., Navathe, S.: An efficient algorithm for mining association rules in large databases. In: Proceedings of the 21st International Conference on Very Large Databases (VLDB). pp. 432–444 (1995)
13. Brin, S., Motwani, R., Ullman, J.D.: Dynamic itemset counting and implication rules for market basket analysis. In: SIGMOD'97, pp. 255–264 (1997)
14. Agrawal, R., Imieliński, T., Swami, A.: Mining association rules between sets of items in large databases. In: Proceedings of the 1993 ACM SIGMOD International Conference on Management of Data, pp. 207–216 (1993)
15. Park, J.S., Chen, M.S., Yu, P.S.: An effective hash-based algorithm for mining association rules. ACM SIGMOD Rec. **24**(2), 175–186 (1995)
16. Toivonen, H.: Sampling large databases for association rules. Vldb. **96**, 134–145 (1996)

Bridge Structure Health Assessment Algorithm Based on Heterogeneous Sensor Data Fusion Simulation

Jianxun Zhang(✉)

Pingdingshan Longze Highway Engineering Co., Ltd., Pingdingshan 467000, Henan, China
2546582645@qq.com

Abstract. The conventional health assessment of bridge structures mainly uses physical data simulation analysis to obtain the assessment results, ignoring the impact of data calibration time differences on the assessment results, resulting in high absolute residual values of the assessment results. Therefore, a bridge structure health assessment algorithm based on heterogeneous sensor data fusion simulation is proposed. Calculate the covariance and convert it into a time state vector, jointly calibrate the data time, perform decision level data fusion, use the fused data to establish a bridge structure simulation model, set bridge structure health status evaluation indicators and perform consistency judgment, calculate the composite weight regeneration evaluation algorithm, and obtain the bridge structure health status evaluation results. The experimental results show that the evaluation results obtained after the application of the proposed algorithm exhibit a low absolute residual value, with a mean of only 2.87, indicating high accuracy and meeting the practical needs of bridge structure operation and maintenance.

Keywords: Bridge structure · Heterogeneous sensor data fusion · Structural health status · Status assessment · Evaluation algorithm · Bridge condition assessment

1 Introduction

Bridges are very important facilities in the field of transportation infrastructure [1]. The water and land transportation functions played by bridges greatly alleviate the traffic pressure on the road surface, thereby promoting the formation and development of towns and driving economic and social development. In order to meet practical usage needs, evaluating the safety of bridge structures has become a focus of attention in the field. Many researchers have proposed methods for evaluating bridge structures.

Reference [2] established an evaluation system by analyzing bridge construction standards, collected real data, and used machine learning algorithms to analyze the state of bridge components. A nonlinear mapping model was established to find the optimal learning curve, but the overall evaluation method was not effective. Reference [3] constructed a digital twin system for bridges based on measured traffic flow loads and

G. Jeon and X. Kong (Eds.): ACIIBD 2024, LNICST 616, pp. 51–57, 2025.
https://doi.org/10.1007/978-3-031-91853-7_6

cable tension detection systems. Monte Carlo simulation technology was used to analyze the ultimate state of bridges, but the evaluation efficiency was relatively low. Based on the data of cable-stayed bridge structures, reference [4] established a comprehensive evaluation system with 3A index as the core by monitoring and analyzing the amplitude offset, mean fluctuation, and anomaly of upper structural components. However, the accuracy of the evaluation needs to be improved. Reference [5] evaluates the health status of bridges by comparing the equivalent modal damping ratio of bridges under different modes with the damping ratio at design time, but the applicability of this method is limited on different bridges. The bridge damage identification method based on combined load response feature fusion proposed in reference [6] has high accuracy and strong noise resistance, but the data acquisition and processing process is complex.

Considering that the bridge structure evaluation method proposed in the above literature cannot meet the practical needs of bridge and operation maintenance work, this study proposes a bridge structure health status evaluation algorithm using heterogeneous sensor data fusion simulation as a means.

2 Design of Algorithm for Evaluating the Health Status of Bridge Structures

2.1 Bridge Structure Data Processing

Install different types of sensors around the bridge structural components to be evaluated. According to the functions of various components in the bridge structure, displacement sensors are installed in the pier foundation, main arch ribs, and columns to collect deflection, temperature sensors are installed in steel truss beams and Y-shaped steel structures to collect temperature, acceleration sensors are installed in steel truss beams and steel box arch ribs to collect acceleration, and suspension rod tension sensors are installed in suspension rods and tie rods to collect cable force.

The data measured by the sensor is heterogeneous, and based on this, the sensor data is fused and processed. Calibrate the time of different sensors jointly. Calibrate individual sensors based on their reference times [7]. Analyze the state vector at the time of sensor acquisition and calculate the covariance. The calculation method of covariance is shown in formula (1).

$$P_{\hat{k}} = AP_k + Q \tag{1}$$

In the formula, k and $\hat{k}$ represent prior and posterior values, respectively, $P_{\hat{k}}$ represents posterior covariance, A represents state transition matrix, P_k represents prior covariance, and Q represents process noise.

As shown in formula (1), the posterior value is predicted by the prior value, and through multiple corrections, it is transformed into a time state vector. As shown in formula (2) [8].

$$\partial = (1 - P_{\hat{k}})^a \tag{2}$$

In the formula, ∂ represents the time state vector, and a represents the filtering gain parameter.

Through the above process, different sensors are time calibrated. After the time calibration of sensors is completed, the data information collected by different sensors will be processed by heterogeneous fusion at a unified time. Calculate the fusion probability of decision results for each heterogeneous data, as shown in formula (3) [9, 10].

$$p(S_{P_k}^1 \middle| S_{P_k}^2) = \frac{\prod_{S=1}^{n} p(s_i | s_j)}{\sum_{S=1}^{n} p(s_i)} \tag{3}$$

In the formula, p represents the fusion probability, S represents the sensor, s_i and s_j represent different data under the same sensor, and n represents the number of sensors.

By fusing heterogeneous sensor data and extracting similar features [11, 12], preprocessing and conducting finite element analysis, a digital model of the bridge is established, and the health status of the bridge is evaluated using MATLAB simulation software.

2.2 Health Assessment of Bridge Structures

Using the bridge structure data fusion simulation model obtained above, set the health status evaluation indicators for the main structure of the bridge, as shown in Table 1.

Table 1. Health status evaluation indicators of bridge main structure

Serial Number	Main structure of bridge	Evaluation indicators
1#	Pier and abutment foundation	1. Cracks $\leq$ 0.2 mm 2. Acceleration < 0.1 g 3. Amplitude $\leq$ 5 mm
2#	Mainarchrib	1. Arch axis linearity $\leq$ L/1000 2. Section stress $\leq$ 24 MPa 3. Cracks $\leq$ 0.2mm 4. Deflection $\leq$ L/400 5. Natural frequency 2.85 Hz–3.15 Hz 6. Acceleration < 0.1 g 7. Amplitude $\leq$ 5 mm 8. Dynamic strain $\leq$ 900 $\mu\,\varepsilon$
3#	Longitudinal beam	1. Section stress $\leq$ 22.5 MPa 2. Cracks $\leq$ 0.2 mm 3. Deflection $\leq$ 83.3 mm
4#	Column	1. Section stress $\leq$ 18.7 MPa 2. Cracks $\leq$ 0.2 mm 3. Deflection $\leq$ 18.75 mm

According to the set health assessment indicators for bridge structures, calculate the composite weights of each indicator [13, 14], which can be expressed as:

$$\omega = \sum_{i=1}^{m} \frac{\gamma_i \cdot v_i}{v_i} \tag{4}$$

In the formula, ω represents the composite weight of the indicators, m represents the evaluation indicators, γ_i represents the weight of each indicator i, and v_i represents the scoring value of indicator i.

Among them, the setting of γ value is modified based on the importance scale of actual bridge structural components [15].

The health status of the bridge structure is divided into different levels based on the rating values output by the algorithm. A score of 90–100 indicates a good condition; 75–90 points are considered minor defects; 60–75 points are classified as moderate defects; A score of 30–60 indicates a significant defect; 0–30 points are classified as severe defects.

3 Experiment

3.1 Experimental Preparation

Design comparative experiments to conduct feasibility testing and effectiveness verification of the proposed bridge structure health assessment algorithm in this study.

Taking a certain highway bridge as an example, conduct a health status assessment on the bridge. The main bridge structure of this highway bridge is a large-span bridge, with a total length of 1260 m, an arch section width of 2.3 m, a main bridge section width of 42 m, a vertical rise of 60.23 m, a suspension rod spacing of 9.5 m, a rise to span ratio of 1:5.7, a main arch rib inclination angle of 10.67°, and a single track weight of 3520 kN.

Set up data monitoring points at various parts of the bridge to collect current data information. The sensor arrangement of each structural component is shown in Table 2.

In Table 3, the main bridge steel truss beam and suspension rod are divided into different sections, and the division of each section is shown in Fig. 1.

Install different types of sensors at the positions shown in Fig. 1 to collect real-time data information of the bridge structure.

Set the algorithm parameters applied in this experiment: total feature count of 100, sample feature count of 0.5, minimum leaf node count of 50, maximum depth of 100, and sample weight of 2. During the iteration process of the algorithm, outliers can be effectively eliminated and the accuracy of the algorithm can be improved.

Based on this, the health status of the bridge structure is evaluated, and the rating values of each component are calculated using formula (4). Each structure is calculated 10 times, and the average value is calculated to obtain the rating values of each component, effectively reducing errors and ensuring the accuracy of the evaluation. The results are shown in Table 3.

According to Table 3, the evaluation results obtained by the method proposed in this article are highly consistent with the actual health status of the bridge structure, and have a certain degree of credibility.

Table 2. Layout of bridge sensor points

Bridge structure	Sensor	Quantity (pieces)
Abutment	Displacement sensor	4
Pier	Displacement sensor	12
Main Bridge Sect. 1	Displacement sensor	3
Main Bridge Sect. 2	Temperature sensor	3
Main Bridge Sect. 3	Acceleration transducer	3
Main Bridge Sect. 4	temperature sensor	3
Main Bridge Sect. 5	Displacement sensor	3
Cross section of suspension rod1	Suspension rod tension sensor	2
Cross section of suspension rod2	Suspension rod tension sensor	2
Cross section of suspension rod3	Suspension rod tension sensor	2
Support	Pressure sensitive sensor	4

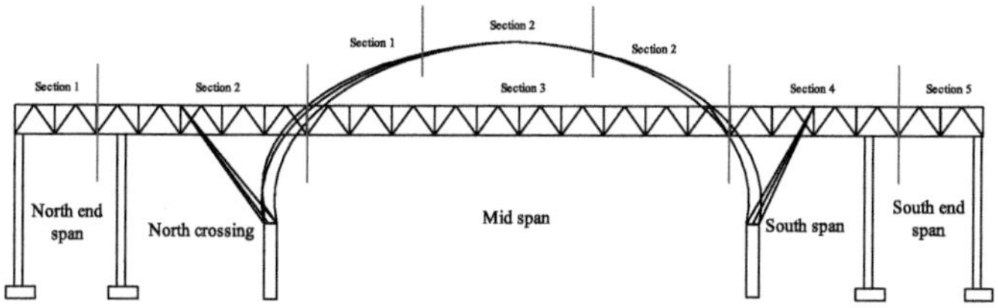

Fig. 1. Cross section division of sensor deployment

Table 3. Health status assessment results of bridge main structure

Structural components	Average component score	Health status
Y-shaped steel structure	75	Minor defects
Steel box arch support	62	Moderate defect
Steel truss beam	75	Minor defects
Boom	79	Minor defects
Tie rod	37	Large defect
Expansion joint	87	Minor defects
Suspension cable	82	Minor defects
Bridge deck	53	Large defect
Bridge deck beam	67	Moderate defect

3.2 Result Evaluation Indicators

The bridge structure evaluation method proposed in references [2] and [3] was used as a comparative method for this evaluation experiment. And use absolute residual value as the evaluation indicator to horizontally compare and analyze the evaluation results of the three evaluation methods. The lower the value of the absolute residual value, the higher the consistency between the evaluation results obtained by this method and the true health status results, and the better the evaluation performance. The result is shown in Fig. 2.

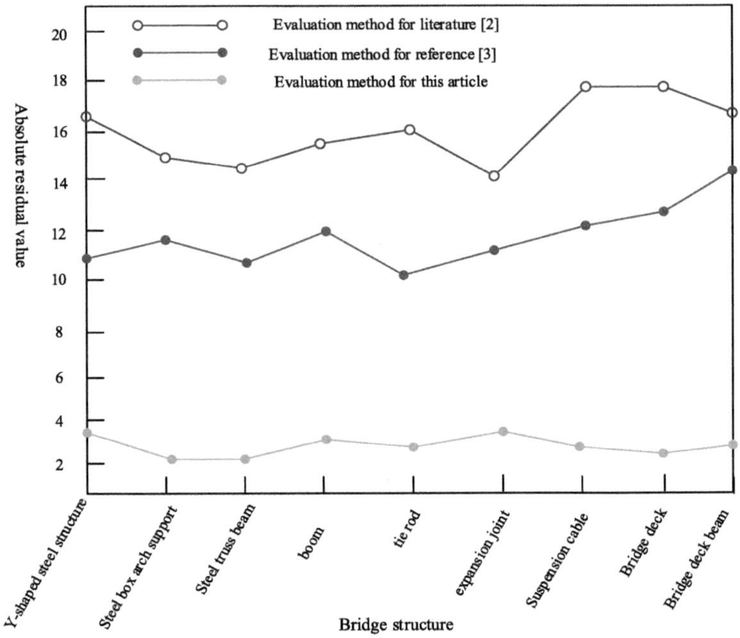

Fig. 2. Absolute residual value of evaluation results

From Fig. 2, it can be seen that the evaluation results obtained by the method proposed in this article show a relatively low absolute residual value. The average absolute residual value of the health status evaluation results of various components of the bridge structure is 2.87, while the evaluation results obtained by the methods proposed in references [2] and [3] show an average absolute residual value of 15.88 and 11.96, respectively. By comparing the evaluation results of the three methods horizontally, it can be seen that the method proposed in this article can effectively reduce the absolute residual value of the evaluation results.

From this experimental result, it can be seen that the proposed method for assessing the health status of bridge structures exhibits a lower absolute residual value, which can effectively improve the accuracy of the evaluation results. Applied in the practical work of bridge structure construction and maintenance, it can obtain reliable evaluation results and have higher practical application value.

4 Conclusion

In order to accurately identify safety hazards in bridge structures, it is crucial to accurately evaluate the health status of bridge structures. This study proposes a bridge structure health assessment algorithm based on heterogeneous sensor data fusion simulation. Through experiments, it is known that the evaluation results obtained based on the proposed method exhibit lower absolute residual values, indicating that this study can effectively improve the accuracy of the health status evaluation results of bridge structures.

References

1. Xu, W., Qian, Y., Zhang, F., et al.: Study on partial safety factors for assessment of existing bridges based on reliability theory considering verification loads. J. Southeast Univ. (Nat. Sci. Ed.) **52**(02), 222–228 (2022)
2. Qiao, P., Liang, Z., Xu, K., et al.: Evaluation of technical condition of medium and small span bridge based on machine learning. J. Chang'an Univ. (Nat. Sci. Ed.) **41**(06), 39–52 (2021)
3. Xu, Z., Song, J., Zhang, X., et al.: Application of digital twin system considering measured traffic flow in structural state evaluation of cable-stayed bridge. J. Highway Transp. Res. Dev. **40**(06), 93–102 (2023)
4. Qin, J., Fang, S., Zhang, W., et al.: Superstructure state evaluation of cable-stayed bridge using 3A indicators. J. Civ. Environ. Eng. **44**(03), 71–78 (2022)
5. Hua, X., Li, P., Wang, T., et al.: Study of modal damping characteristics of steel-concrete composite bridges based on strain energy analysis. China J. Highway Transp. **36**(07), 47–55 (2023)
6. Zhong, Y., Zhang, H., Zhang, G., et al.: An intelligent damage identification method based on response feature fusion under combined loadings for bridge structures. J. Exp. Mechan. **38**(02), 151–164 (2023)
7. Raleigh, L.Z.: Application of structural health monitoring system in dynamic response analysis of large-span bridges under earthquake action. Build. Struct. **53**(S1), 2122–2126 (2023)
8. Yang, M., Mei, D., Huo, X., et al.: Research on refined calculation method of time-dependent prestress loss of bridge structure. World Bridges **51**(S1), 152–158 (2023)
9. Qin, F.: Research on the finite element model correction method for small and medium span bridges considering structural damage. Highway **68**(03), 176–181 (2023)
10. Nie, Z., Cai, W., Xia, Z., et al.: Bridge structural damage identification based on high-resolution ECOMAC. Vibr. Test. Diagn. **43**(01), 179–187+205
11. Ma, Y., Wang, M., Jia, D., et al.: Design of heterogeneous sensor information fusion display based on QT. Foreign Electron. Measur. Technol. **41**(09), 176–182 (2022)
12. Li, S., Yue, L., Li, Z., et al.: Sea surface target recognition and tracking system for unmanned surface vehicle with heterogeneous sensors. Chinese J. Ship Res. **16**(S1), 131–137 (2021)
13. Wenya, Jia, J., Zhang, B., et al.: A heterogeneous sensor allocation method based on particle swarm optimization algorithm. Fire Control Command Control **46** (04), 54–58 (2021)
14. Zhu, J., Gu, W., Luo, D., et al.: Multi-sensor data fusion based on PSO optimized BP neural network. China Measur. Test. Technol. **48**(08), 94–100 (2022)
15. Ma, K., Zhang, Y., Wu, H., et al.: Research on model test of cutting landslide based on multi-sensor data fusion analysis. J. Disaster Prevent. Mitigat. Eng. **42**(04), 653–663 (2022)

The Volatility Prediction and Option Pricing Model of Correction Bias

Ruijie Cui(✉)

School of Science, Dalian Minzu University, Dalian 116000, Liaoning, China
crj13610984579@163.com

Abstract. Aiming at the problem of prediction bias in GARCH volatility model, the prediction bias of GARCH model is corrected based on improved recurrent neural network. This paper constructs an option pricing model with predictable volatility in an effort to reduce the estimation error of the model and attain more accurate volatility estimation. The experimental results indicate that the model can significantly update the prediction accuracy of GARCH model. Based on this, the stochastic differential and martingale methods are used to get the option pricing formula under the risk neutral condition. To certain degree, it eliminate the shortcoming of the traditional artificial designed volatility model, which can only estimate the parameters by using the option market price. The numerical results show that the precision of the system is ideal, and it is found that the B-S formula for 50ETF stock option pricing is often less than the option pricing in this paper.

Keywords: component · Recurrent neural network · GARCH model · option pricing system · risk-neutral pricing

1 Introduction

Options are financial derivatives with risks. Reasonable pricing of options helps investors to make rational decisions and avoid possible risks effectively, while reasonable pricing of options by financial regulatory authorities is conducive to the standardized management of the market and the maintenance of financial market stability. The traditional parameterized option pricing model has strict economic logic, but its pricing effect is not good when the assumption is too ideal. Traditional methods of forecasting volatility are widely used in financial markets, but they can only capture linear relationships in time series because they need to satisfy strict parameter constraints, however, it can not deal with the non-linear relationship in time series, so it performs poorly in prediction. There is no ideal volatility model to price and hedge financial derivatives, so it is important for investors and regulators to predict market volatility accurately.

Santamaría et al. [1] combined a hybrid genetic algorithm (HGA) with Support vector machine regression (SVRGBC) to predict volatility and found that the SVRGBC model outperformed the GARCH model in predicting volatility. Hamid et al. [2] used ANN model to forecast the volatility of S & p 500 index futures option, and found

G. Jeon and X. Kong (Eds.): ACIIBD 2024, LNICST 616, pp. 58–67, 2025.
https://doi.org/10.1007/978-3-031-91853-7_7

that the forecasting effect of Ann was better than that of implied volatility. In 2022, Wei Heping [3] proposed a FNN-AE-LSTM-RV-mv model, which uses fully connected neural networks and long-term and short-term memory neural networks, at the same time, the volatility of Shanghai Stock Exchange 50ETF factors to build a model to study the volatility. Nevertheless, owing to the nonlinear and long-term dependence of time series data, it is difficult for these models to effectively capture the key time series characteristics [4, 5]. In addition, due to the problems of gradient vanishing and gradient descends, these models can not make use of the past information to predict as full as possible. In order to improve the accuracy and reliability of volatility prediction, it is necessary to filter the high-dimensional features of the data set and find the financial factors that have the greatest influence on the volatility data, so as to construct a volatility model that can better capture the time series.

The GARCH model has a strong memory for early volatility and can not respond quickly to new volatility changes, whereas the machine learning model has fewer data requirements and assumptions and can handle diverse and complex structure data [6]. In order to make up the advantages of GARCH model and machine learning algorithm, the volatility deviation captured by GARCH model is corrected by using machine learning algorithm [7–9], a hybrid model based on machine learning deviation correction is proposed to predict the volatility of stock market, which can improve the accuracy of the volatility of stock market return and provide a new way to predict the volatility of financial market, by using stochastic differential and martingale methods, the European option pricing formula is obtained under the conditions of risk neutrality and predictable volatility.

2 Related Theories

2.1 RNN Recurrent Neural Network

The traditional Feedforward neural network is a network of input, hidden, and output layers [10]. RNN network expansion, then parameters w, u, v are shared, but the traditional neural network is not. As shown in Fig. 1. And in the gradient descent algorithm, the output of each step depends not only on the network of the current step, but also on the state of the previous steps, after linear transformation between the state vector of the last timestamp and the input of the current timestamp, a new state vector is obtained by the activation function through the transfer of the network cells between the recurrent neural networks, its output depends not only on the current input, but also on the current memory. This memory is sequence-to-sequence, where the present moment is affected by the previous moment, such as a change in the stock market [11].

2.2 GARCH Model

The traditional time series model assumes that the variance of financial assets is constant, which is contrary to the actual characteristics of price fluctuation in financial market. Therefore, Engle (1982) [12] uses the square of the residual lag term and its weight to

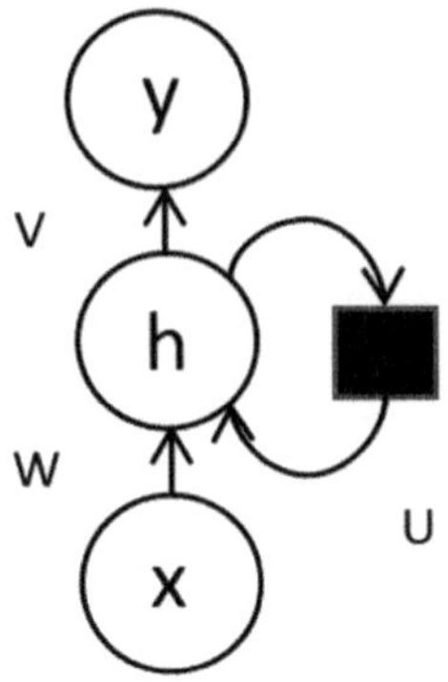

Fig. 1. Circular neural network

describe the variance of the subject volatility again, and then constructs the classical autoregressive conditional heteroskedasticity (ARCH) model:

$$\sigma_t^2 = \alpha_0 + \sum_{j=1}^{q} \sigma_{t-j}^2 + \sum_{i=1}^{p} \varepsilon_{t-i}^2 \tag{1}$$

Among them ε_{t-i}^2 is called the ARCH term, σ_{t-j}^2 is called the GARCH term, q, p are orders of two terms. GARCH model can reduce the computation of ARCH model, and can describe ARCH model with high order hysteresis by low order hysteresis.

3 Option Pricing Model Construction

3.1 Volatility NRNN Model Design

The volatility NRNN model designed in this paper is composed of two sub-networks, one is a few recurrent neural networks predicting implied volatility, the other is a weight neural network. In order to improve the learning ability of input data to historical data, the previous acquired information or knowledge in machine learning system decreases with time, f_t it mainly determines how much data from the previous moment can be preserved to the current moment, x_t is the input vector at t, The t component of the input sequence x, x_t is transformed linearly by multiplying the weight matrix w. h_{t-1} holds the data at time $t-1$, h_{t-1} is transformed linearly by multiplying the weight matrix W. The two types of data information are added and then activated by the *sigmoid* function, The numerical result is therefore between f_t (0, 1), as shown in Formula (2).

$$f_t = sigmoid(W_{fh}h_{t-1} + W_{fx}x_t + b_f). \tag{2}$$

The NRNN hidden layer i_t introduces the entire stream information c_{t-1} up to the last moment,At the same time, the introduction of c_{t-1} will also have an impact on the retention of current time data.NRNN contains r hidden layers, The weights between the $i-1$ hidden layer and the i hidden layer are expressed in matrix w_i, bwu_i is the offset vector of the i hidden layer, $2 \leq i \leq r$, The forward propagation part of NRNN contains k fully connected layers, which essentially constitutes a multi-layer Feedforward neural network. h_i^r represents the i node of the r fully connected layer, the number of nodes in the r full connection layer is n_r, The link between nodes of the fully connected r and $r+1$ layers has a weight of U_{ij}^r and an offset of bwu_r, Then c_t is the data information preserved from the beginning to the present moment, as shown in Formula (3).

$$c_t = f_t c_{t-1}. \tag{3}$$

h_j controls how much of the data retained from the beginning to the current moment can be passed on to the next hiding layer, as shown in Formula (4).

$$h_j^{r+1} = sigmoid(\sum_{i=1}^{n_r} U_{ij}^r X_i^r + W_{ij}^r h_i^r + c_t + bwu_r^j). \tag{4}$$

$$j = 1, \ldots n_{r+1}, r = 1, \ldots k - 2. \tag{5}$$

Y_{ij} represents the output information, as shown in Formula (6).

$$Y_{ij} = sigmoid(\sum_{i=1}^{k-1} h_j^{k-1} V_I^{K-1} + bwu_r^j). \tag{6}$$

The entire network has two parameter matrices $[W_{fh}, W_{fx}]$ and $[W_{ih}, W_{ix}]$, as $W_{i=}[W_{fh}, W_{fx}]$ and $W = [W_f, W_i]$, And L_t is the loss corresponding to code h_t. It is recorded as the total loss of L, and the W derivative of L is shown in Formula (7).

$$\frac{\partial L}{\partial W} = \sum_{t=1}^{T} \frac{\partial L_t}{\partial W}. \tag{7}$$

The recurrent neural network updates the W parameter by Formula (8).

$$W = W - \frac{\partial L_t}{\partial W}. \tag{8}$$

Remember that $t = d$, $\partial L_t = \partial W$ in Formula (7) can be rewritten as (9).

$$\frac{\partial L_d}{\partial W} = \frac{\partial L_d}{\partial h_d} \frac{\partial h_d}{\partial c_d} \frac{\partial c_d}{\partial c_{d-1}} \cdots \frac{\partial c_2}{\partial c_1} \frac{\partial c_1}{\partial W}. \tag{9}$$

3.2 Option Pricing Formula of Base Volatility NRNN Model

[Theorem 1] let the price of the asset at time t satisfy the Stochastic differential equation:

$$dS(t) = \alpha(t)S(t)dt + \sigma(t)S(t)dW(t), 0s \leqslant t \leqslant T \tag{10}$$

The volatility function given in this paper is

$$\sigma^2(t) = f_{\text{NRNN}}\left(\left\{\sigma^2(u)\right\}_{t,q}; \theta\right) \tag{11}$$

In this paper (9) and (10) using Ito-Doeblin Formula:

$$S(T) = S(t)\exp\left\{\int_t^T \left(\alpha(u)\frac{1}{2}f_{\text{NRNN}}\left(\left\{\sigma^2(u)\right\}_{t,q}; \theta\right)\right)du + \int_t^T \sqrt{f_{\text{NRNN}}\left(\left\{\sigma^2(u)\right\}_{t,q}; \theta\right)}dW(u)\right\} \tag{12}$$

According to the principle of risk-free arbitrage, when the portfolio is not affected by randomness, is only related to time t, and there is no external risk factor, the return rate of the portfolio is equal to the risk-free interest rate R, thus, in the case of continuous compounding, the following Stochastic differential equation are obtained

$$dS(t) = rS(t)dt + \sqrt{f_{\text{NRNN}}\left(\left\{\sigma^2(u)\right\}_{t,q}; \theta\right)}S(t)d\tilde{W}(t) \tag{13}$$

According to the risk-neutral pricing formula, the price at the moment of maturity of a European call with a strike price of T is:

$$d_+\left(t, T, S(t), \left\{\sigma^2(u)\right\}_{t,q}, K; \theta\right) = \frac{1}{\sqrt{\int_t^T f_{\text{NRNN}}\left(\left\{\sigma^2(u)\right\}_{t,q}; \theta\right)du}}$$

$$\left[\ln\frac{S(t)}{K} + r(T-t) + \frac{1}{2}\int_t^T f_{\text{NRNN}}\left(\left\{\sigma^2(u)\right\}_{t,q}; \theta\right)du\right] \tag{14}$$

$$d_-\left(t, T, S(t), \left\{\sigma^2(u)\right\}_{t,q}, K; \theta\right) = \frac{1}{\sqrt{\int_t^T f_{\text{NRNN}}\left(\left\{\sigma^2(u)\right\}_{t,q}; \theta\right)du}}$$

$$\left[\ln\frac{S(t)}{K} + r(T-t) - \frac{1}{2}\int_t^T f_{\text{NRNN}}\left(\left\{\sigma^2(u)\right\}_{t,q}; \theta\right)du\right] \tag{15}$$

Among them $\left\{\sigma^2(u)\right\}_{t,q} = \left\{\sigma^2(t-q), \cdots \sigma^2(t-1)\right\}$, $W(t)$ is Brownian motion in probability space $(\Omega, \mathcal{F}, \mathbb{P})$, $\mathcal{F}(t)$ is about the watershed of the Brownian motion, $\alpha(t)$ is the average rate of return, $\sigma(t)$ is the volatility of asset prices at time t, q is the lag period of the volatility time series and θ is the parameter of the volatility NRNN model, $\sqrt{\int_t^T f_{\text{NRNN}}\left(\left\{\sigma^2(u)\right\}_{t,q}; \theta\right)du}$ is the integral of the wave-law function.

4 Computational Experiments

In this paper, we take the five indexes of the real volatility of the lagged period, the extreme difference of the closing price of the Shanghai Stock Exchange 50 ETF, the closing price, the trading volume and the real volatility of the lagged two periods as the input of each machine learning algorithm, the current volatility bias is taken as the output, and the modified bias sequence is obtained by training the prediction bias of GARCH model with the improved machine learning algorithm, taking the sum of the predicted values of GARCH model and GARCH model as the predicted values of the final GARCH model volatility, the GARCH volatility prediction model based on machine learning bias correction is constructed, and the main flow is shown in Fig. 2.

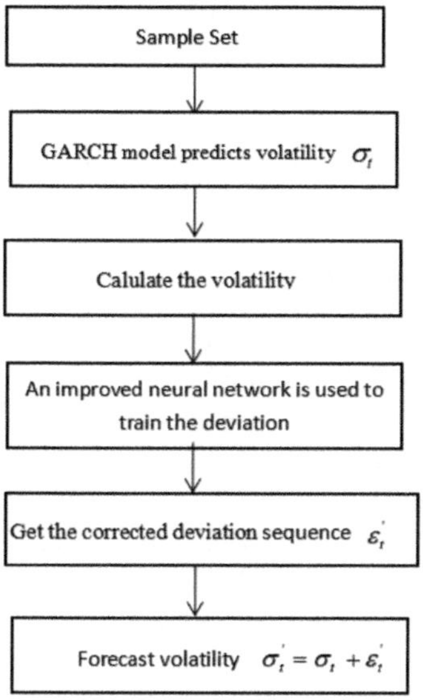

Fig. 2. Volatility bias correction prediction model

4.1 Data Description

In order to test the actual effect of this system, using the Shanghai Exchange 50ETF stock options, select the daily closing price of the China AMC Shanghai 50 ETF (2005-2-23 to 2023-1-16, Data Source Choice Financial Terminal), a sample of 4357 trading days was used to calculate historical volatility. The sample set is divided into training set and test set, and the rolling time window prediction method is used to predict the model.

$$\frac{S_{\mathrm{i}+1} - S_i}{S_i} = R_i. \tag{16}$$

$$\bar{R} = \frac{1}{M}\sum_{i=1}^{M} R_i. \tag{17}$$

In Process: R_i is a logarithmic rate of return, $\overline{R}$ is the average of log returns over n days, Based on relevant academic experience, n is always 22.Volatility deviation is defined as the difference between the predicted result of the GARCH model and the true volatility, i.e.:

$$RV = \sqrt{\frac{1}{M-1}\sum_{i=1}^{M}(R_i - \bar{R})^2}. \tag{18}$$

Volatility bias is defined as the difference between the predicted results of the GARCH model and the true, real volatility, i.e.:

$$\varepsilon_t = \sigma_t - RV_t \tag{19}$$

where S_i is the closing price of the marked asset price on day i and σ is the daily volatility.

4.2 Analysis of Experimental Results

The prediction bias of GARCH model is trained by machine learning algorithm, and the network is trained by BP algorithm of adaptive learning rate. On the training set, the average absolute error of model generalization is 1.3472×10^{-4}, which is close to the actual result. On the test set, the average absolute error of model generalization is 5.3486×10^{-5}, and the generalization ability is ideal. The accuracy of the test set is higher because the training set contains "Big Jump" data, and the generalization ability of the model needs to be improved. Figure 3 shows how the GARCH model generalization results compare with the actual situation on the test set.

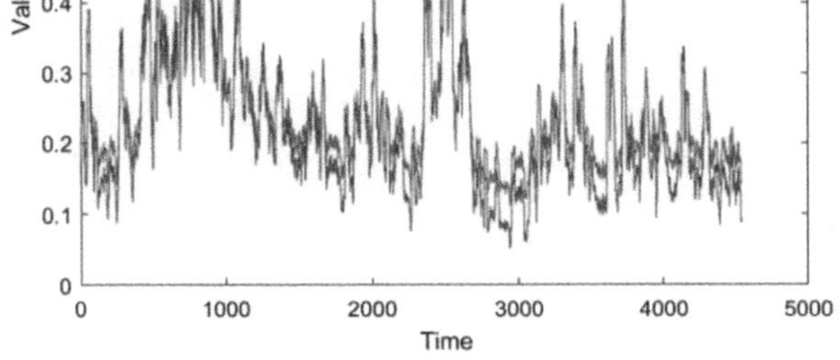

Fig. 3. Comparison of model generalization to reality

Using 4537 individuals as the training set, the curve of RBM reconstruction error from input layer to middle layer after 5000 iterations is shown in Fig. 4.

The reconstruction error is defined as:

$$RE = \text{mean}\left(\left|neg(p_v^l) - v^l\right|\right). \tag{20}$$

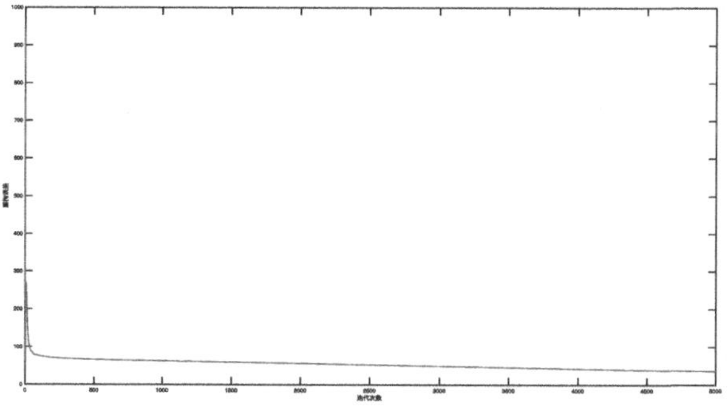

Fig. 4. Error curve

Table 1 shows the comparison of the results of 50ETF stock option pricing between the system and B-S formula. The daily closing prices of 50 ETF funds in bull market, bear market and horizontal market are S1 = 0.9134, S2 = 2.6500 and S3 = 2.3500, respectively, options with 6,34 and 69 days remaining. The risk-free yield was 0.02383. By comparison, we find that the pricing of 50ETF stock option by B-S formula is lower than that of the system in this paper, and the degree of the pricing will increase with the increase of the time remaining from the maturity date. And the degree of underpricing will be larger for the average option, and smaller for the deep real and imaginary options, K that increases with time.

5 Conclusions

Volatility plays a major role in the pricing of financial derivatives, asset allocation and risk management [13, 14]. With the integration of the global financial market, the demand for the measurement of volatility is also increasing, so a reasonable model is constructed to measure the volatility of financial assets and reveal the nature of financial market volatility, it is of great theoretical and practical value to the investors and the financial supervision department. In this paper, the return of Shanghai Stock Exchange 50ETF is taken as the research object, and a GARCH volatility prediction model based on machine learning deviation correction is proposed, under the assumption of predictable volatility, a non-parametric volatility model is established, based on which an option pricing formula is obtained and an option pricing system is constructed. Based on this, a closed-form solution is obtained which can be used to determine the theoretical price. The numerical results show that the model can describe the movement of the volatility of asset price well, but there is a big deviation in describing the "Big jump", which is one of the main problems in the future. In this paper, the option pricing problem affected by non-parametric volatility model is studied.

The system can provide a method for investors to calculate the option price and control the investment risk.

Table 1. The comparison of option pricing and B-S formula under the volatility deviation positive model

S	K	Price (T = 6)			Price (T = 34)			Price (T = 64)		
		B-S	Theorem	Bias	B-S	Theorem	Bias	B-S	Theorem	Bias
		1			1			1		
S1	0.6234	0.2902	0.2902	0	0.2914	0.2914	0	0.2928	0.2933	−0.0005
	0.6734	0.2403	0.2403	0	0.2415	0.2417	−0.0002	0.2430	0.2449	−0.0019
	0.7234	0.1903	0.1903	0	0.1916	0.1915	−0.0009	0.1933	0.1985	−0.0053
	0.7734	0.1403	0.1403	0	0.1417	0.1454	−0.0037	0.1435	0.1555	−0.0121
	0.8234	0.0903	0.0906	−0.0003	0.0918	0.1027	−0.0109	0.0937	0.1175	−0.0238
	0.8734	0.0403	0.0449	−0.0046	0.0419	0.0670	−0.0251	0.0439	0.0854	-0.0415
	0.9234	0	0.0141	−0.0104	0.0003	0.0402	−0.0339	0.0026	0.0598	−0.0571
	0.9734	0	0.0024	−0.0024	0	0.0221	−0.0221	0	0.0403	−0.0403
SS2	2.3423	0.3086	0.3129	−0.0043	0.3129	0.3129	0	0.3182	0.3145	−0.0038
	2.4423	0.2087	0.2131	−0.0045	0.2131	0.2135	−0.0004	0.2187	0.2206	−0.0020
	2.4923	0.1587	0.1632	−0.0046	0.1632	0.1650	−0.0018	0.1689	0.1775	−0.0086
	2.5423	0.1087	0.1133	−0.0046	0.1133	0.1192	−0.0059	0.1191	0.1382	−0.0191
	2.5923	0.0587	0.0635	−0.0048	0.0634	0.0789	−0.0154	0.0697	0.1039	−0.0342
	2.6423	0.0087	0.0182	−0.0094	0.0159	0.0468	−0.0309	0.0263	0.0751	−0.0488
	2.6923	0	0.0008	−0.0008	0.0001	0.0244	−0.0243	0.0042	0.0521	−0.0479
	2.7423	0	0	0	0	0.0111	−0.0111	0.0002	0.0346	−0.0345
SS3	2.1069	0.2439	0.2445	−0.0006	0.2478	0.2479	−0.0002	0.2526	0.2544	−0.0018
	2.1569	0.1939	0.1960	−00021	0.1979	0.1987	−0.0008	0.2028	0.2073	−0.0045
	2.2069	0.1440	0.1498	−0.0058	0.1480	0.1509	−0.0029	0.1530	0.1629	−0.0099
	2.2569	0.0940	0.1078	−0.0138	0.0981	0.1067	−0.0086	0.1032	0.1227	−0.0195
	2.3069	0.0440	0.0722	−0.0282	0.0482	0.0689	−0.0207	0.0535	0.0880	−0.0345
	2.3569	0	0.0446	−0.0445	0.0032	0.0399	−0.0367	0.0100	0.0559	−0.0498
	2.4069	0	0.0251	−0.0251	0	0.0204	−0.0204	0.0001	0.0384	−0.0383
	2.4569	0	0.0129	−0.0129	0	0.0092	−0.0092	0	0.0232	−0.0232

References

1. Santamaría-Bonfil, G., Frausto-Solís, J., Vázquez-Rodarte, I.: Volatility forecasting using support vector regression and a hybrid genetic algorithm. Comput. Econ. **45**(1), 111–133 (2015)
2. Hamid, S.A., Iqbal, Z.: Using neural networks for forecasting volatility of S&P 500 Index futures prices. J. Bus. Res. **57**(10), 1116–1125 (2004)
3. Wei, H.: Research on realized volatility based on multivariable-hybrid neural network model. Shandong University (2022). 10.27,272/d.CNKI.GSHDU.2022.002227
4. Li, P., Yang, J., Lin, Y.: Parameter estimation and application of stochastic volatility model based on fast mean regression. Oper. Res. Manag. **29**(2), 137–143 (2020)
5. Brandt, M., Jones, C.: Volatility forecasting with range-based EGARCH models. Working Paper, Fuqua (2002)
6. Liu, Y.: Novel volatility forecasting using deep learning-long short term memory recurrent neural networks. Expert Syst. Appl. **132**, 99–109 (2019)

7. Cao, W., Ren, S.: A study on forecasting the fluctuation of RMB exchange rate based on the mixed model of LSTM and GARCH family. Comput. Appl. Res. **37**(s1), 79–82 (2020)
8. Lu, X., Que, D., Cao, G.: Volatility forecast based on the hybrid artificial neural network and GARCH-type models. Procedia Comput. Sci. **91**, 1044–1049 (2016)
9. Li, X., Xu, J., Wang, Y., etc.: Establishment and application of BP artificial neural network adaptive learning algorithm. Theory Pract. Syst. Eng. **24**(5), 1–8
10. Chen, W., Xu, G.: Research on the prediction accuracy of stock market volatility based on deep learning and stock forum data. Managing the world (2018)
11. Yū, H.: Financial time series analysis and prediction based on deep learning and network big data. Sichuan: Sichuan Normal University (2022). Fan Tienteng, Wu Peiyu. Study on volatility of Shanghai 50ETF based on GARCH (1,1) model [J] . Time Finance, 2017, no
12. Engle, R.F.: Autoregressive conditional heteroscedasticity with estimates of the variance of United Kingdom inflations. Econometrica **50**(4), 987–1007 (1982)
13. Alghalith, M.: Pricing the American options using the BlackScholes pricing for-mula. Physica A Stat. Mechan. Appl. **507**, 443–445 (2018)
14. Yang, T., South, H.: Study on volatility of stock returns based on Garch Family Model. Manag. Inform. China **24**(23), 128–129 (2021)

Attention-Based Link Prediction with Contextualized Self-supervision

YunHai Gao[1], GuiYun Zhang[1(✉)], Jian Zhang[2], and YueXiu Zhang[3]

[1] Tianjin Normal University, Tianjin 300000, China
dyxy1999@126.com
[2] Nankai University, Tianjin 300000, China
[3] Tianjin Agricultural University, Tianjin 300000, China

Abstract. This paper presents a novel link prediction model named ACSS (Attention-based Link Prediction with Contextualized Self-Supervision), which integrates attention mechanisms to address challenges faced by traditional link prediction algorithms. Conventional methods are limited by sparse links, noisy node attributes, and dynamic variations. The ACSS model combines contextualized self-supervised learning with attention mechanisms to better capture critical relationships between nodes and improve link prediction performance. Our model employs end-to-end training, optimizing it by simultaneously learning link prediction and self-supervised learning tasks. Experimental show that the ACSS model surpasses traditional and inductive link prediction tasks across multiple real-world benchmark networks. Furthermore, the ACSS model exhibits strong robustness and scalability when handling node attribute noise and large-scale networks.

Keywords: Link prediction · attention mechanism · contextualized self-supervised learning · graph neural networks

1 Introduction

Link prediction in graph-structured data is a crucial task with broad applications and challenges. In many real-world scenarios, it is often necessary to predict unknown network links to better understand and analyze complex network structures, or to implement important applications such as social network analysis, recommendation systems, bioinformatics, and network security.

Link prediction faces numerous challenges. Networks typically exhibit complex topological structures and dynamic evolution patterns. Traditional methods often overlook these complexities, focusing solely on local similarity between nodes while ignoring global topological information. Moreover, real-world network datasets often suffer from data sparsity and incomplete labels. Traditional methods require a large number of known link samples for model training, but it is challenging to obtain a sufficient quantity of high-quality label data in practical applications.

G. Jeon and X. Kong (Eds.): ACIIBD 2024, LNICST 616, pp. 68–75, 2025.
https://doi.org/10.1007/978-3-031-91853-7_8

To address these challenges, we propose a novel link prediction model, the Attention-based Link Prediction with Contextualized Self-Supervision (ACSS). This model integrates attention mechanisms with contextual self-supervised learning to improve the accuracy and generalization capability of link prediction. The ACSS model utilizes attention to dynamically learn the associations between nodes and adaptively weight the importance of nodes and edges, capturing both global structures and local features of the network. The introduced contextual self-supervised learning mechanism leverages the topological structure and node attribute information of the network, enhancing robustness in the face of data sparsity and incomplete labels.

Our contributions include: Application of Attention Mechanism: By using attention mechanisms to model contextual information in graph data, we better capture the relationships and importance between nodes, improving the accuracy and effectiveness of link prediction. Proposal of the ACSS Model: We propose the ACSS model, which combines attention mechanisms with self-supervised learning to provide an effective solution to the link prediction problem. Self-supervised learning is utilized to more accurately predict link relationships in graph data.

The structure of this paper is as follows: Sect. 2 reviews traditional prediction methods and discusses the integration of self-supervised learning and attention mechanisms with graph neural networks in the context of graph data. Section 3 provides a detailed explanation of the principles and overall structure of the ACSS model. Section 4 describes the experimental setup, datasets, evaluation metrics, and comparative experiments. Section 5 summarizes the research findings and outlines future directions.

2 Related Work

2.1 Traditional Link Prediction Methods

Traditional link prediction methods rely heavily on the structural attributes of graphs, where links are predicted by assessing the similarity between nodes. Shortest Path Index [1]: Evaluates similarity based on the minimum path length between two nodes. Adamic-Adar Index [2]: Considers the significance of neighboring nodes by using the inverse logarithm of shared neighbors as a measure of similarity.

These methods perform well on small and medium-sized graphs but may have limitations on large-scale and complex graph data, prompting researchers to explore link prediction methods based on deep learning and graph neural networks.

2.2 Self-supervised Learning for Link Prediction

Self-supervised learning, which has achieved success in computer vision and natural language processing, is gradually being applied to graph-structured data. The idea is to design effective self-supervised tasks that generate supervision signals from unlabeled data, avoiding costly manual labeling processes. Self-supervised learning has shown significant results and great potential in solving link prediction tasks.

Self-supervised tasks leverage the inherent structures and patterns in data, such as node neighborhood information, network topology, or node attributes, to generate meaningful supervision signals, reducing dependence on labeled data and improving model generalization and robustness. In graph-structured data, self-supervised learning can better adapt to the characteristics of real-world data, enhancing the model's representation and feature-capturing capabilities. Attri2Vec [3]: Integrates node attribute information into embeddings, combining node structure and attribute information for training. VGAE [4]: A variational autoencoder-based unsupervised node embedding and link prediction model, trained by reconstructing the graph adjacency matrix. CSSL [5]: A contextualized self-supervised learning model that uses contextual information in graph data for self-supervised learning.

2.3 Attention Mechanisms for Link Prediction

Attention mechanisms have great potential in link prediction, simulating the human focus mechanism in information processing to dynamically learn and focus on different parts of the graph. Graph Attention Networks (GATs) [6]: Introduce attention mechanisms in graph neural networks, focusing on important nodes and links to improve link prediction performance. Att-ConvLSTM [7]: A dynamic link prediction method employing hierarchical attention to learn diverse information, integrating recurrent neural networks with temporal attention to capture evolving patterns.

Existing research demonstrates the various advantages of attention mechanisms in processing graph-structured data. They dynamically allocate attention weights, better capture the relationships between nodes and edges, enhance model focus on important information, and improve model interpretability and explainability.

3 Model

3.1 Related Definitions

Assume a network $G = (V; E; X)$ where V is the set of nodes, E is the set of edges, and $X \in R^{m \times |V|}$ is the matrix of node attributes, representing the information for each node. The edge set E is split into a training set E_{tr} and a test set E_{te}. The objective is to develop a link prediction model using the known training edge set E_{te} and node properties X or X_{tr}.

3.2 ACSS Model Introduction

The ACSS model integrates attention mechanisms and contextual self-supervised learning. It employs a random walk strategy similar to node2vec to sample node sequences in the graph, thereby obtaining contextual information about the nodes. An attention layer is added to the model, focusing on both node attributes and structural information, which enhances the model's ability to represent node embeddings. This effectively improves the model's capacity to capture critical information.

The design and operational principles of the ACSS model include obtaining contextual information, embedding contextual information, self-supervised learning based on attention mechanisms, and the final link prediction process. The ACSS model can be divided into two main components: contextual information embedding and self-supervised learning based on attention mechanisms (see Fig. 1).

We used a random walk method to generate a sequence of context nodes for each node to capture local structural features. We refer to the Node2vec model, using the same sampling distribution method, and adjust the hyperparameters to balance the depth and breadth of their walk to better obtain the context information $C(v)$.This sequence includes neighboring nodes with strong relationships with the node.

After obtaining the context node sequence C(v), we aggregate the features of these nodes to generate the context embedding representation of the target node v:

$$h_v^{context} = \frac{1}{|C(v)|} \sum_{u \in C(v)} X_u \tag{1}$$

where $h_v^{context}$ represents the context embedding of node v, C(v) is the sequence of the context nodes of the node v, and X_u is the feature representation of context nodes u. The contextual information embedding part obtains more representative node embeddings, which are used for the subsequent attention mechanism-based self-supervised learning part. The idea of the attention mechanism is to assign different weights to neighbor nodes, highlighting the features of neighbors that significantly influence the current node, thereby generating more accurate node representations. For node v and its neighbor nodes N(v), we calculate their attention weights α_{vu}:

$$\alpha_{vu} = \frac{exp(LeakyReLU(a^T[Wh_v \| Wh_u]))}{\sum_{k \in N(v)} exp(LeakyReLU(a^T[Wh_v \| Wh_k]))} \tag{2}$$

where $N(v)$ denotes the set of all neighbouring nodes of node v, h_v and h_u are the feature representations of node v and its neighbouring node u, respectively, W is the weight matrix to be learnt, and a is the parameter of the attention mechanism Vector, $\|$ denotes the vector splice operation.The attention weights are used to generate new feature representations by weighted summing of the features of the neighbor nodes:

$$h_{v'} = \sigma\left(\sum_{u \in N(v)} \alpha_{vu} Wh_u\right) \tag{3}$$

where σ is a non-linear activation function, such as ReLU.

To optimize the node representations, we need an appropriate loss function. The contrastive loss function, initially proposed by Yann LeCun et al. [8], measures the similarity between paired samples. It distinguishes positive and negative sample pairs by minimizing positive pair distances and maximizing negative pair distances.

Calculate contrast loss for each node and sum the contrast loss of all nodes to obtain the overall loss function:

$$\mathcal{L}_{acss} = \sum_{i \in V} \ell_i = \sum_{i \in V}\left[-log\frac{exp\left(\frac{h_i \cdot h_i^+}{\tau}\right)}{\sum_{j \in N(i)} exp\left(\frac{h_i \cdot h_j}{\tau}\right)}\right] \tag{4}$$

The ACSS model optimizes node feature representations layer by layer through contextual information embedding and attention mechanism-based self-supervised learning. It also uses a contrastive loss function to optimize the overall model, allowing the model to better learn the relationships between nodes.

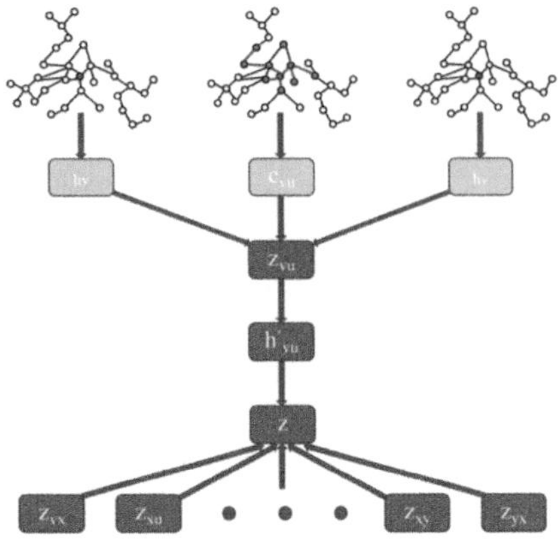

Fig. 1. ACCS model structure

4 Experiments

4.1 Dataset Description

To verify the effectiveness of the proposed ACSS model, experiments were conducted on five widely used graph datasets: three citation networks (Cora, Citeseer), two web networks (WebKB, Wiki), and one social network (Facebook). These datasets are summarized in Table 1.

Table 1. Dataset statistics.

Dataset	Cora	Citeseer	WebKB	Wiki	Facebook
Type	Citation network	Citation network	webpage networks	webpage networks	Social Network
Classes	7	6	7	–	–
Nodes	2708	3327	877	2405	4039
Features	1433	3703	1703	4973	1406
Edges	5429	4732	1480	24357	88234

4.2 Experimental Setup

During data preprocessing, each dataset's edge set was split into training, validation, and test sets in an 85%, 5%, and 10% ratio, respectively. For experimental fairness, identical training parameters were used for all models: a learning rate of 0.01, a batch size of 20, and 200 training epochs.

To evaluate link prediction models' performance, the Area Under the ROC Curve (AUC) was employed as the metric. AUC measures a binary classification model's performance, with values ranging from 0 to 1; a higher AUC signifies better performance.

$$AUC = \int_0^1 TPR(FPR^{-1}(t))dt \quad (5)$$

4.3 Results and Analysis

We compared the performance of various baseline models on the five datasets. The models compared are: DIG [9]: A graph neural network enhanced with dual-view information, capturing both topological relationships and attribute similarities between nodes. GIM [10]: Maximizes mutual information between local and global node representations to generate more informative node embeddings. DEAL [11]: Combines edge attribute information and node features for deep edge attribute learning, used for link prediction tasks.The experimental results are summarized in Table 2, with the best results in bold and the second-best underlined.

Table 2. Area under the ROC curve (AUC) for link prediction tasks.

Method	Cora	Citeseer	WebKB	Wiki	Facebook
Attri2Vec	<u>0.923 ± 0.002</u>	<u>0.959 ± 0.003</u>	<u>0.913 ± 0.008</u>	0.870 ± 0.007	0.912 ± 0.005
DIG	0.879 ± 0.004	0.936 ± 0.004	0.869 ± 0.019	0.847 ± 0.010	0.909 ± 0.004
GIM	0.889 ± 0.008	0.924 ± 0.008	0.812 ± 0.019	0.829 ± 0.009	<u>0.926 ± 0.005</u>
VGAE	0.899 ± 0.003	0.913 ± 0.006	0.897 ± 0.015	<u>0.904 ± 0.004</u>	**0.945 ± 0.005**
DEAL	0.822 ± 0.010	0.874 ± 0.009	0.861 ± 0.022	0.819 ± 0.010	0.862 ± 0.004
ACSS	**0.942 ± 0.001**	**0.980 ± 0.001**	**0.949 ± 0.002**	**0.920 ± 0.003**	0.917 ± 0.001

By comparing the results of the ACSS model with those of the baseline models, we can observe that ACSS achieves better performance on four out of the five datasets. This suggests that our model's attention mechanism effectively captures relational features between nodes and generalizes well across different datasets.

The ACSS model did not perform well on the Facebook dataset. The Facebook dataset, as a social network dataset, has a clear community structure, which makes the global properties of the graph more complex. Limited inter-community links make capturing global structural information challenging, indicating a need for improvement in handling datasets with unique structures.

4.4 Ablation Study

To validate the effectiveness of the attention mechanism in the ACSS model, we conducted ablation experiments comparing it with two structured CSSL models. The experimental results are presented in Table 3.

Table 3. Comparison of Ablation Experiment Results.

Method	Cora	Citeseer	WebKB	Wiki	Facebook
CSSL_Joint	0.936 ± 0.002	0.977 ± 0.002	0.941 ± 0.005	0.916 ± 0.002	0.922 ± 0.003
CSSL_Pretr	0.932 ± 0.003	0.972 ± 0.002	**0.958 ± 0.005**	0.912 ± 0.002	**0.928 ± 0.002**
ACSS	**0.942 ± 0.001**	**0.980 ± 0.001**	0.949 ± 0.002	**0.920 ± 0.002**	0.917 ± 0.001

The CSSL_Joint and CSSL_Pretr models employed neighborhood joint and neighborhood pre-training strategies, respectively. Each strategy showed advantages on different datasets. The attention mechanism calculates weights based on node correlation, allowing the model to focus on important neighborhood information. In our ablation experiments, the ACSS model with attention mechanism demonstrated superior performance on most datasets, highlighting the effectiveness of the attention mechanism. Significant performance improvements were observed in predicting Cora, Citeseer, and Wiki datasets. However, the ACSS model performed poorly on the Facebook dataset, likely due to the distinctive community structure inherent in social network datasets, which constrained the performance of the attention mechanism.

5 Conclusion

This study proposes ACSS, an attention-based context self-supervised link prediction model. The attention mechanism allows the model to capture node relationships and importance more effectively. Experiments on five popular datasets showed that ACSS outperformed others, proving the attention mechanism's effectiveness. However, the ACSS model also has some limitations, showing decreased performance on certain specific datasets. The model's performance is limited when dealing with datasets with special structures. In future research, we aim to explore more effective attention mechanisms to improve adaptability to various types of datasets.

Acknowledgments. This work was supported by Tianjin Research Innovation Project for Postgraduate Students (2022SKYZ266).

References

1. Wasserman, S., Faust, K.: Social Network Analysis: Methods and Applications. Cambridge University Press (1994)
2. Adamic, L.A., Adar, E.: Friends and neighbors on the web. Soc. Networks **25**(3), 211–230 (2003)
3. Zhang, M., Chen, Y.: Link prediction based on graph neural networks. In: Advances in Neural Information Processing Systems, vol. 31 (2018)
4. Kipf, T.N., Welling, M.: Variational graph auto-encoders. arXiv preprint arXiv:1611.07308 (2016)

5. Zhang, D., Yin, J., Philip, S.Y.: Link prediction with contextualized self-supervision. IEEE Trans. Knowl. Data Eng. **35**(7), 7138–7151 (2022)
6. Veličković, P., Cucurull, G., Casanova, A., Romero, A., Liò, P., Bengio, Y.: Graph attention networks. arXiv preprint arXiv:1710.10903 (2017)
7. Ni, X., Zhao, Y., Yao, Y.: Dynamic heterogeneous link prediction based on hierarchical attention model. In: Proceedings of the 8th International Conference on Cyber Security and Information Engineering, pp. 111–115 (2023)
8. LeCun, Y., Chopra, S., Hadsell, R., Ranzato, M., Huang, F., et al.: A tutorial on energy-based learning. Predicting Struct. Data **1**(0) (2006)
9. Zhou, J., et al.: Graph neural networks: a review of methods and applications. AI Open **1**, 57–81 (2020)
10. Veličković, P., Fedus, W., Hamilton, W.L., Liò, P., Bengio, Y., Hjelm, R.D.: Deep graph infomax. arXiv preprint arXiv:1809.10341 (2018)
11. Hao, Y., Cao, X., Fang, Y., Xie, X., Wang, S.: Inductive link prediction for nodes having only attribute information. arXiv preprint arXiv:2007.08053 (2020)

Intelligent System Design and Information Processing

Lightweight MES System Research and Practice

A Medium-sized Discrete Manufacturing Enterprise as An Example

Zhihong Ren(✉) and Di Zhang

School of Information Engineering, Xi'an Eurasia University, Xi'an 710000, Shaanxi, China
{renzhihong,zhangdi}@eurasia.edu

Abstract. "Industry 4.0" has triggered a major change in the field of intelligent manufacturing, MES (Manufacturing Execution System) is the core software system in the field of intelligent manufacturing, and it is an important hand to promote intelligent enterprise management. The popularization and promotion of MES system is a necessary way to promote China to become a "manufacturing power". This paper takes the construction and implementation of "lightweight" MES for medium-sized discrete manufacturing enterprises as the research background, takes "cost reduction and efficiency" as the goal, and relies on the platform of "school-enterprise joint laboratory" to construct and implement a MES system adapted to the current situation of the enterprise. With the goal of "cost reduction and efficiency", relying on the platform of "school-enterprise joint laboratory", we have constructed and implemented a "lightweight" MES solution adapted to the current situation of the enterprise, and achieved good application results.

Keywords: MES · smart manufacturing · digital enterprise management · enterprise informatization

1 Introduction

The concept of MES was first proposed by AMR (Advanced Manufacturing Research) in 1990, which is defined as "a shop floor oriented management information system located between the upper level planning management system and the lower level industrial control". It provides operators/managers with plan execution, tracking, and current status of all resources (people, equipment, materials, customer requirements, etc.). Its functions include production order management, manufacturing data management, and program scheduling management, Production scheduling management Inventory management, quality management, Human Resource Management, work center/equipment management, tools and fixtures management, cost management, project kanban management, production process control, bottom data integration and analysis, top data integration and decomposition and other management modules, to create a solid, reliable, comprehensive and feasible manufacturing collaborative management platform for enterprises [5].

G. Jeon and X. Kong (Eds.): ACIIBD 2024, LNICST 616, pp. 79–87, 2025.
https://doi.org/10.1007/978-3-031-91853-7_9

2013 Hannover Messe Germany officially launched the concept of "Industry 4.0" [1], the core of which is the use of information technology to promote industrial change, and promote human beings to enter the age of intelligence [2]. "Industry 4.0" is based on the concept of intelligent manufacturing as the leading fourth industrial revolution or revolutionary production methods, aiming to transform the manufacturing industry to intelligence by making full use of information and communication technology and cyberspace virtual system - information physical system (Cyber-Physical System) [3]. In May 2015, China's State Council formally issued the "Made in China 2025" [4], deployed to comprehensively push forward the implementation of the Manufacturing Power Strategy[4]. Industry 4.0 has entered the China-Germany cooperation New era, China and Germany signed the Program of Action for Sino-German Cooperation which clearly states that the digitalization of industrial production is "Industry 4.0" and is of great significance for the future development of the German-Chinese economy.

Industry 4.0 provides a reference standard for the informatization of manufacturing industry, and MES has become an innovative technology for realizing Industry 4.0. With "Smart Manufacturing" as the core, MES has become a support platform for lean production, and plays a core role in reducing production cost and improving product quality. Parametric, platform-based MES products, powerful and flexible, can be quickly adapted to the customer, with the improvement of customer management level continues to evolve, and become a "smart manufacturing" platform in the era of Industry 4.0.

Zhang [11] proposed a lightweight model-driven MES simulation framework based on probabilistic finite automata. The framework realizes highly flexible production process simulation through configuration files and generates production data.

2 Analysis of the Development Status of MES System at Home and Abroad

In November 1990, the U.S. Advanced Manufacturing Research Center AMR (Advanced Manufacturing Research) has proposed MES (Manufacturing Execution System In 1997, MESA proposed MES functional components and integration model, including 11 functions, while providing that as long as one or more of the 11 functions, also belongs to the MES series of single-function products [5]. Into the twenty-first century, the application of foreign well-known enterprises MES system has become a common phenomenon, on behalf of manufacturers Siemens, SAP, GE Fanuc, Gallup, etc., these global head of the enterprise with its technological superiority and huge market share in the field of MES in the leading position, relying on the industrial Internet [7], the Internet of Things, intelligent perception, cloud computing and other modern information technology, and continue to Relying on modern information technologies such as industrial Internet, Internet of Things, intelligent perception, cloud computing, etc., they continue to promote the deepening research, popularization and promotion of intelligent manufacturing technologies. The earliest MES in China was introduced from Siemens at the early stage of the construction of Baosteel in 1980s, and China's industrial informatization started a new research boom, and in recent years, the "One Step Cloud" MES system of HANDY, the C-H Event-based Manufacturing Execution System, the Foshan Real-time Manufacturing Execution System, and the Foshan Real-time Manufacturing Execution System are all in the leading position in the field of MES. In recent years, HANDE Information "One Step Cloud" MES system, Zhonghao real-time manufacturing execution system (C-H Event-based Manufacturing Execution System), Foshan Zhongyuan Technology Co [8].

In recent years, with the proposal of new production modes such as JIT (Just In Time) and BTO (order-oriented production), as well as the higher requirements of customers and markets for product quality, MES has only been rediscovered and emphasized [6]. Meanwhile, after the bursting of the network economy bubble, enterprises began to realize that they have to improve their competitiveness from the most basic production management, i.e., only by taking out the data and information from the product level (the basic automation level), passing through the operation and control level, and delivering them to the management level, and realizing the integration of enterprise information through the continuous flow of information can the enterprise be invincible in the increasingly fierce competition. MES has been rapidly and widely applied in foreign countries. Manufacturing Execution System (MES)[1] aims to enhance the execution capability of enterprises with irreplaceable functions. Enterprises in the process industry under the competitive environment should distinguish the objectives and roles of different manufacturing management systems, clearly define the positioning of MES in the integrated system, pay attention to the accuracy and timeliness of the information, standardize the process, utilize the tools, and manage the innovations, and analyze their own execution capability based on the MES maturity model [3], and follow the information Integration, Transaction Processing, Manufacturing Intelligence three-stage step-by-step implementation of the MES system, in order to give full play to the role of enterprise informatization, improve the competitiveness of the enterprise, and bring the expected benefits to the enterprise [9].

3 Design and Implementation of Lightweight MES System

3.1 Lightweight MES System Design Ideas

Modular Design. The MES system is split into several functional modules, each of which is responsible for different production management tasks, such as order data, process parameters, equipment data, etc.. Enterprises can select the appropriate module for deployment according to their own needs.

Simplified User Interface. Intuitive and simple user interface design enables non-professionals to get started quickly. Through the graphical interface and wizard-type operation process, reduce the user's learning costs.

Rapid Deployment and Implementation. Design lightweight MES systems with rapid deployment and implementation in mind. With pre-configured settings and templates, the implementation cycle is reduced, accelerating the pace of enterprise digital transformation.

Data Security and Backup. Ensure data security by providing data encryption, user rights management and data backup functions to prevent data loss or leakage.

Scalability. The design should take into account the scalability of the system, with the development of enterprises and changes in production requirements, the system can be easily expanded and upgraded.

3.2 Analysis of Enterprise Status and System Requirements

The enterprise is a typical discrete manufacturing enterprise, product characteristics: multi-species, small batch, high value-added, product planning based on market demand, randomness, 0 inventory, delivery pressure. The enterprise deployed SAP ERP system one year ago for enterprise business integration management. Lightweight MES development is required to realize the following functions:

Docking with SAP ERP production management module (PP), real-time access to order data, process parameters, equipment data, material BOM data from ERP, and at the same time need to access to the basic database of employee information, working hours calculation logic, as a source of input data to the MES system. After the closing of a production order in the MES system, docking with the SAP ERP system, the completed order data written to the PP module to complete the order closed-loop management process. PP module to complete the closed-loop management process of the order.

Workshop dispatch function realization. Through the product process flow, the products in the order are divided into work processes, and according to the equipment and operator information, the operator dispatch work order is generated and can be printed out.

Receiving material management. Combined with the product material BOM data, to generate the operation of the worker's receipt, as the warehouse issued material vouchers.

Labor reporting management. The system integrates the function of enterprise WeChat, and workers complete the operation of reporting work through enterprise WeChat.

Workshop display board function. Show real-time information of production orders through the big screen, which is convenient for real-time production scheduling in the workshop.

Statistical analysis function. Statistical analysis of the daily production completion of the workshop, generate statistical charts, and compared with the production plan statistical task completion rate. Statistical charts are pushed to the workshop director, factory director and company senior leaders through WeChat, providing different granularity of statistical charts for different levels of leadership.

System background function. Suitable for system administrator roles for system basic information import and maintenance, running log management and other system management work.

3.3 Lightweight MES System Architecture Design

The lightweight MES system architecture is shown in Fig. 1. It includes network device layer, data layer, business logic layer and data presentation layer [12]. The network equipment layer includes devices and sensors, application gateways, SAP ERP cloud servers, communication links and TCP/IP networks, constituting the physical topology of the system and the infrastructure for the operation of the entire system. The data layer includes the basic database, ERP intermediate database, MES database and data analysis database. The basic database is used to store basic information such as employees, equipment, enterprise organization structure, etc. The ERP intermediate database is the storage buffer between the MES system and the ERP database, which is used for data exchange between the two systems to avoid the data risk caused by direct operation of the ERP database. The business logic layer includes SAP ERP production order import, SAP ERP BOM/process data import, enterprise WeChat interface management, real-time data monitoring, basic information maintenance, work assignment management, material management, workshop display board management, work report management, statistical reports and log records. Business logic layer is the core process of system management, based on in-depth user calls, and its algorithmic advantages and disadvantages have a significant impact on the system's operational efficiency. The data display layer includes the PC side of work assignment management (administrator), the mobile side of work assignment/reporting management (worker), the mobile side of statistical report, monitoring information, the large screen side of workshop display board, the workplace process card display board, the PC side of statistical report (business) and early warning information. The data display layer is developed based on the user's work scenarios, and it needs to take into full consideration of the data requirements and statistical data granularity of different work scenarios and the convenience of accessing data in different application scenarios. Consider the convenience of different application scenarios to access the data, such as what scenarios are presented through the mobile terminal, what scenarios are presented through the PC terminal, the user must be the center of the data display design.

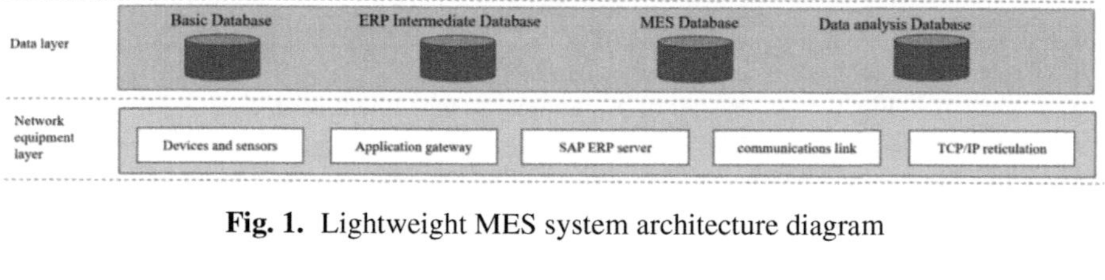

Fig. 1. Lightweight MES system architecture diagram

3.4 Lightweight MES System Realization

The overall system implementation framework adopts Microsoft's.NET Framework, the database supporting Microsoft's SQL SERVER, and the business logic is realized through C# language programming [10]. On the mobile side, the enterprise has introduced the enterprise WeChat platform, and the system realizes the display of user data on

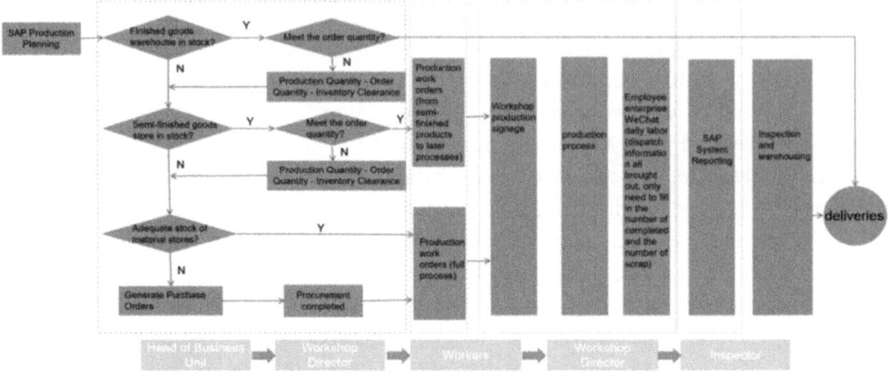

Fig. 2. Flowchart

the mobile side by calling the enterprise WeChat development interface, which improves the ease of use of the system [11]. The running interface of workshop director assignment management is shown in Fig. 2. The data analysis running interface is shown in Fig. 3, the enterprise WeChat user login interface is shown in Fig. 4, and the worker's enterprise WeChat report interface is shown in Fig. 5, Production plan allocation management is shown in Fig. 6.

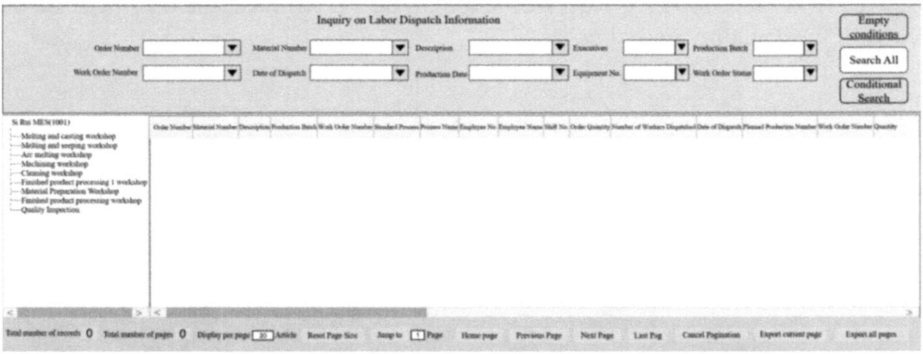

Fig. 3. Screenshot of assignment management operation

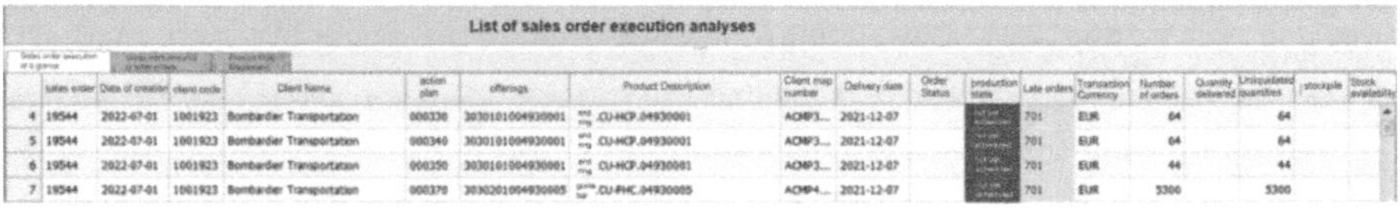

Fig. 4. Screenshot of data analysis run

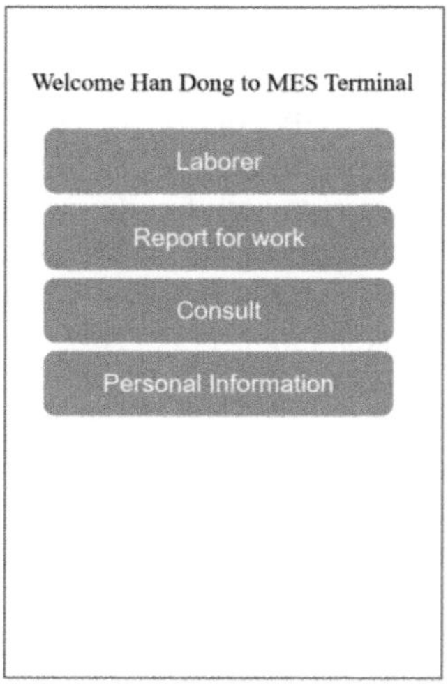

Fig. 5. Screenshot of mobile enterprise WeChat login

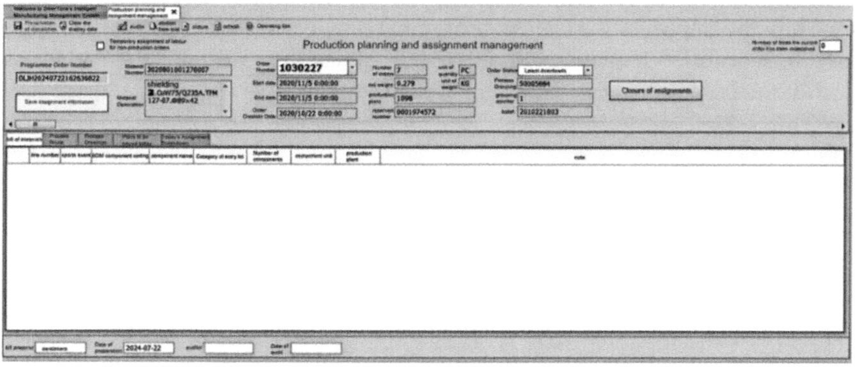

Fig. 6. Production Plan Assignment Management

3.5 Major Problems and Countermeasures for Project Promotion

Three months after the project went live, the following problems occurred successively, and solutions were given after research to ensure that the system was continuously optimized and upgraded in an iterative manner.

System Operational Efficiency Issues. At the early stage of the system on-line, there were problems of slow running speed and low efficiency, which affected the dispatching. Solution strategy: As the dispatching business involves SAP order data, product process design, BOM data, etc., which need to be obtained directly from SAP ERP server, large data volume, complex data logic, and will involve SAP ERP security mechanism, for this reason, the system set up an ERP intermediate database, for order data, process data, BOM data preprocessing, to improve the system operation efficiency. The operation efficiency of the system is improved.

SAP ERP System Data Write Problems. The SAP ERP system deployed by this enterprise is a cloud solution, and the system security mechanism is particularly stringent, and the writing of system data is strictly controlled. The MES system developed in this project belongs to a plug-in system of SAP ERP system, and the order, process and BOM data are obtained from SAP ERP system, and after the completion of the production order, it is necessary to complete the shut-off operation in SAP ERP system, and the system process forms a closed-loop to ensure the accuracy of the data in SAP ERP. The problem is solved by consulting the technical engineers of SAP.

Management Issues. The on-line of the new system has completely changed the original workflow of the workshop management, involving multiple roles from the workshop director, technicians, post workers, etc. Due to the varying levels of informatization literacy and cognitive level of these personnel, coupled with the system on-line debugging stage, the productivity of the post workers will have a certain impact on the workers' productivity, and there are problems such as unfamiliarity of the workers to report their work through the cell phone terminal. Therefore, at the initial stage of system on-line, some employees have resistance, which affects the application of the system. Solution strategy: Strengthen user education and training, improve personnel informatization literacy, improve staff awareness of the importance of informatization work, while taking into account the impact of the project promotion on the personal interests of employees, remove the worries of employees, mobilize the enthusiasm of employees, and the strong support of the user is the key to the success of the project.

4 Summary and Outlook

Lightweight MES system development and deployment after more than a year, which experienced a difficult work history, the enterprise management team, project development team, business operations team is an important experience, summarize the project experience and lessons can be summarized in a few points:

4.1 Perceptions of Digital Transformation in the Enterprise

Enterprise digital transformation is the inevitable trend of the times, is the enterprise in the future competition in the invincible important guarantee, the project once started, must adhere to go on, there is no way back. These perceptions must be established from the top management of the enterprise to the shop floor grass-roots staff awareness. Therefore, enterprise digital transformation must be "a handful of projects", must be deployed through the top management of enterprises, comprehensive promotion, full participation, which is an important guarantee for the smooth progress of the project. Through its automated production processes, real-time data monitoring, resource optimization, strict quality control, efficient order management, customer feedback integration, and continuous improvement mechanisms, the MES system significantly increases productivity, reduces production costs, and greatly enhances customer satisfaction, intuitively demonstrating its key role in improving the competitiveness of the manufacturing industry and market responsiveness.

4.2 Full Recognition of the Complexity of the Problem

MES system development is a complex system engineering, involving the enterprise production method face (man, machine, material, method, environment), as well as the enterprise's existing workflow and product characteristics, the deployment of the system is carried out on the basis of the enterprise's production activities can not be interrupted, how to seamlessly switch, how to do without affecting the production of the workshop, resulting in a decline in the workshop production are important issues faced in the process of promoting the project.

References

1. Hu, C., Lu, J., Ben, K., Ca, W., Zhang, L.: Research and application of key technologies for development and integration of multi-tenant discrete MES. In: International Conference on Artificial Intelligence and Electromechanical Automation (2020)
2. Bianchini, A., Savini, I., Andreoni, A., Morolli, M., Solfrini, V.: Manufacturing execution system application within manufacturing small-medium enterprises towards key performance indicators development and their implementation in the production line. Sustainability. **16**, 2974 (2024)
3. Ahuett-Garza, H., Kurfess, T.R.: A brief discussion on the trends of habilitating technologies for Industry 4.0 and Smart manufacturing. Manuf. Lett. **15**, 60–63 (2018)
4. When "Made in China 2025" meets Germany's "Industry 4.0". China government website (2018)
5. Goel, R., Gupta, P.: Robotics and Industry 4.0. In: Nayyar, A., Kumar, A. (eds.) A Roadmap to Industry 4.0: Smart Production, Sharp Business and Sustainable Development; Advances in Science, Technology & Innovation. Springer, Cham (2020)
6. Lin, S., Yan, Z., Wang, J.: Functions and Practices of Manufacturing Execution System (MES). People's Posts and Telecommunications Publishing House (2021)
7. Xu, X., Ding, X.: Exploring the reform of new engineering talent cultivation mode oriented to sustainable competitiveness. China Univ. Educ. (6), 6–10 (2017)
8. Guo, H., Zhang, L., Zeng, Z.: Exploration and practice of software engineering specialization construction under the background of new engineering science-taking Longyan college as an example. J. Longyan College **9**, 119–122 (2020)
9. Guo, X.: Research on talent cultivation mode of computer "race and education fusion" in colleges and universities–Taking Harbin Guangsha College as an example. J. Harbin College **40**(8), 139–141 (2019)
10. Hu, Z., Tang, H.: Discussion on the reform of practice teaching system of computer majors in applied undergraduate colleges. Software Guide **19**(2), 192–194 (2020)
11. Zhang, K., Tian, J., Han, J., Yuan, Q.: A lightweight model-driven MES simulation framework based on probabilistic finite automata. In: International Conference on Intelligent Computing and Signal Processing (ICSP), Xi'an, China, pp. 1093–1096 (2022)
12. Beregi, R., Pedone, G., Háy, B., Váncza, J.: Manufacturing execution system integration through the standardization of a common service model for cyber-physical production systems. Appl. Sci. **11**(16), 7581 (2021)

A Genetic Deep Reinforcement Learning Approach with Prioritized Experience Replay Strategy to Solve the Influence Maximization Problem on Networks

Zhixi Chen, Junge Wu, Shuai Wang(✉), and Jing Huang

School of Intelligent Systems Engineering, Shenzhen Campus of Sun Yat-Sen University, Shenzhen 518000, Guangdong, China
{chenzhx86,wujg6,huangj525}@mail2.sysu.edu.cn,
wangsh368@mail.sysu.edu.cn

Abstract. The influence maximization problem is a critical task pertaining to social networks, aiming to identifying a set of key individuals who can achieve the maximal information spread through the system. Great attention has been drawn in related studies, and traditional methods often struggle with prohibitive computational efficiency and limited scalability. Therefore, in this paper, a novel approach is developed to solve this problem by leveraging a Prioritized Experience Replay (PER) method within the framework of Deep Reinforcement Learning (DRL), named PER-GDRL. This approach includes a specialized PER method to enhance the learning efficiency and performance of DRL agents. More informative experiences can be prioritized in the training process. Additionally, principles of the PER method are deliberated to optimize the Markov process, making an efficient seed search and selection process. Experimental results demonstrate that PER-GDRL outperforms existing techniques in terms of both performance and computational efficiency. A promising direction is given for applying advanced reinforcement learning techniques to optimization problems related to complex networks.

Keywords: The influence maximization problem · prioritized experience replay · deep reinforcement learning · evolutionary optimization

1 Introduction

The Influence Maximization (IM) is a fundamental problem when analyzing social networks, a set of key individuals are expected to maximize the spread of information across the whole network. This problem has significant applications in various domains, including viral marketing, information dissemination, and public health campaigns. The IM problem is generally modeled as a combinatorial optimization problem, where the objective is to select a limited number of seeds to trigger the maximal influence under certain diffusion models. Additionally, the IM problem has also been widely studied in some directed networks and multilayer networks [1].

G. Jeon and X. Kong (Eds.): ACIIBD 2024, LNICST 616, pp. 88–97, 2025.
https://doi.org/10.1007/978-3-031-91853-7_10

There are many methods to solve the IM problem currently, such as heuristic algorithms like PageRank [2] and IICEA [3]. However, these heuristic algorithms often suffer from high computational complexity and poor scalability. Additionally, some population-based search methods have been applied to the influence maximization problem, including discrete particle swarm optimization [4] and Memetic algorithms [5]. These methods typically involve multiple simulations of the diffusion process to estimate influence spread, which is computationally expensive. To address these challenges, recent research has explored the employment of machine learning techniques, particularly the deep reinforcement learning (DRL) [6]. However, existing algorithms are generally unable to perform efficiently on large-scale networks, and these DRL algorithms usually have low utilization of experiences, leading to problems such as slow model iteration. Therefore, in this paper, a novel approach called PER-GDRL (Prioritized Experience Replay in Genetic Deep Reinforcement Learning) is developed to solve the IM problem. Our method leverages the Prioritized Experience Replay (PER) technique within the framework of Genetic Deep Reinforcement Learning (GDRL). Additionally, the Markov decision process is enhanced by integrating the principles of the PER method, making the seed search and selection process more efficiently. The main contributions of this paper are as follows: (1) We designed a specialized PER method for the IM problem, optimizing the Markov decision process and improving the efficiency of seed selection. (2) We proposed a novel GDRL framework that combines DRL with genetic algorithms, enhancing the exploration and exploitation capabilities of the DRL agent. (3) We conducted extensive experiments to evaluate the performance of our proposed PER-GDRL method. The results show that our method outperforms existing techniques in terms of influence spread and computational efficiency.

2 Related Work

2.1 Propagation Model

In this work, we chose the IC model as the propagation model. The IC model is one of the most commonly used propagation models in social network analysis. The propagation process of the IC model starts with an initial set of active nodes, known as seed nodes. Each activated node has one chance to activate its inactive neighbors. For each activated node u and its inactive neighbors v, there exists a propagation probability p_{uv} that determines the likelihood of u activating node v. If node u successfully activates node v, then v will become activated in the next time step. This process proceeds in discrete time steps. In each step, newly activated nodes attempt to activate their inactive neighbors. The process terminates when no more activations occur.

2.2 Influence Evaluation Method

In social network analysis, evaluating the influence of a seed set is an important research topic. In this work, we adopt a fast influence evaluation method based on the 2-hop neighborhood of seed nodes was proposed in [7]. For a network $G_{in} = \{V, E\}$ with k

seed nodes, the 2-hop influence evaluation method is defined as follows:

$$\hat{\sigma}(S) = \sum_{v_i \in S} \hat{\sigma}(v_i) - \left(\sum_{v_i \in S} \sum_{v_i \in T_i \cap S} p_{ij} \cdot (\sigma^1_{v_j} - p_{ij}) \right) - \sum_{v_i \in S} \sum_{v_j \in T_i \cap S} \sum_{v_k \in T_i \cap S} p_{ij} p_{jk} \quad (1)$$

where, S represents the seed set, and v_i represents a seed node. $\hat{\sigma}(S)$ and $\hat{\sigma}(v_i)$ represent the number of nodes influenced by the seed set S and seed node v_i under the IC model using the 2-hop neighborhood method, respectively. T_i represents the 1-hop neighborhood set of the seed nodes v_i. $\sigma^1_{v_j}$ represents the 1-hop influence of a seed node v_j, calculated as $\sigma^1_{v_j} = 1 + \sum_{v_k \in T_j} p_{kj}$. In this equation, p_{ij} represents the activation probability from an activated node v_i to an inactive node v_j.

2.3 Algorithms for Finding Influential Nodes

Due to the complexity and large scale of social networks, finding the optimal set of seed nodes becomes an NP-hard problem. To address this challenge, researchers have proposed various algorithms and methods, including population-based search methods and reinforcement learning (RL) algorithms related to the method discussed in this study. Population-based learning methods include genetic algorithms, particle swarm optimization [8], etc. In the influence maximization problem, RL can model the node selection process as a Markov Decision Process (MDP) and transform the IM problem into a sequential decision problem [9]. In summary, algorithms based on reinforcement learning and Markov Decision Processes provide new solutions and methods for the influence maximization problem.

3 Methodology of PER-GDRL

This section introduces core modules of PER-GDRL and its overall framework. Four components are included, i.e., the network embedding, the deep reinforcement learning module, the exploration module and genetic operations.

3.1 Algorithm Core Module

Graph Embedding. The effects of different encoding methods on solving the influence maximization problem have been discussed in [10]. Extensive experiments have demonstrated that the Structural Deep Network Embedding (SDNE) encoding method [11, 12] outperforms other encoding methods. Therefore, this work continues to use the SDNE encoding method for network embedding. In the experimental section, we discuss the impact of other encoding methods on our algorithm.

In [11], a semi-supervised deep model for performing network embedding (SDNE) was proposed, and its framework is shown in Fig. 1. SDNE is utilized to represent nodes in the network using low-dimensional vector spaces, thereby extracting latent features and transforming discrete network data into continuous features. Ultimately, this results in a matrix H^D of size $n \times (d + 1)$ as the input for the DQN module.

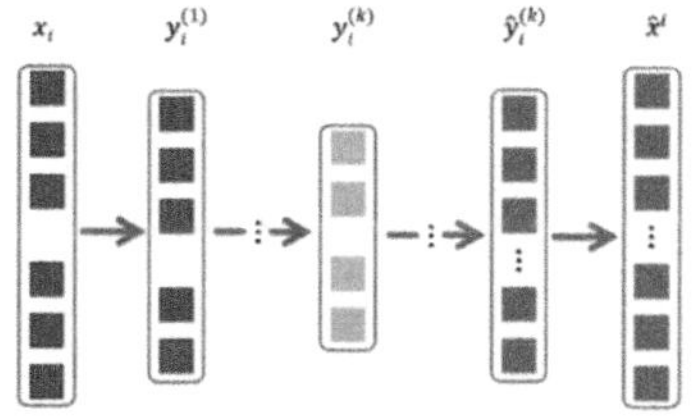

Fig. 1. Diagram of SDNE Encoding and Decoding Structure.

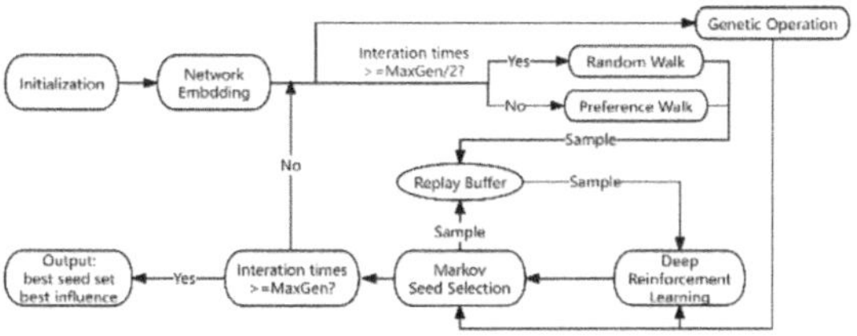

Fig. 2. The Flowchart of PER-GDRL

Deep Reinforcement Learning Module. *Deep Q-Network (DQN).* DQN is an model that combines deep learning and reinforcement learning. The DQN model approximates the Q-value function using a deep neural network, enabling effective policy learning in high-dimensional state spaces. In $Q(S, a)$, S represents the node state, a represents the action of selecting the current node. $Q(S, a)$ is the approximated Q-value function, representing the reward that can be obtained when taking action a. The calculation method of $Q(S, a)$ in this part is $\mathbf{Q}(S, a) = \mathbf{DQN}(W^1, W^2, H^D, S) = \mathbf{ReLU}(\mathbf{Y}(H^D, S) \cdot W^1) \cdot W^2$. Where W^1 and W^2 represent the weight parameters of the first layer and the second layer of the DQN network, respectively. DQN() represents the three-layer DQN network used in this work. $\mathbf{Y}$ is a node feature selection function that can prevent seed nodes from being repeatedly selected. For each node i, $\mathbf{Y}(H_i^D, S_i) = H_i^D \cdot (1 - S_i)$.

Markov Seed Selection. In this study, we employed the Markov seed selection strategy to choose seeds. The core idea of this strategy is to utilize the theory of the Markov decision process (MDP) to optimize seed selection through state transitions and reward mechanisms. Its goal is to select the optimal action in the current state and push $(S_t, a_t, r_t, S_{t+1}, I_{t+1})$ into the DQN's replay buffer. r_t is the reward obtained after taking action $a.I_{t+1}$ represents the influence of the seed set at time $t + 1$ after executing action a at time t.

Continuous DQN Parameter Update. This work designs a Prioritized Experience Replay (PER) mechanism to ensure that more useful samples are obtained for weight updates. In [10], the strategy is to randomly select a batch of samples of size B from the replay buffer for parameter updates each time weights are updated. However, this strategy can lead to high randomness in the selected samples, which may provide insufficient effective information for parameter updates. Therefore, this work designs a PER mechanism to ensure that the samples selected each time are more targeted. Specifically, for each sample, the absolute value of the difference between the maximum influence in replay buffer and their influence is recorded as I_d. After calculating I_d for all samples, I_d is used to calculate the probability C of each sample being selected. Specifically, when the number of samples in the replay buffer is n, the probability C_z of sample z being selected can be expressed as $C_z = I_{zd}/\sum_{j=1,2,\ldots,n} I_{jd}$. Based on the calculated probability C_z for each sample, B samples are selected from the experience pool without replacement, following the calculated probabilities, for parameter gradient updates.

For the B samples selected, the DQN weights (W^1, W^2) for each sample in B are updated using Stochastic Gradient Descent (SGD) by minimizing the loss function L. The formula for the loss function is given in Eq. (2), and the update formulas for the network parameters are given in Eq. (3).

$$L(W^1, W^2) = \underset{C \in B}{\mathrm{E}}[(\sum_{i=0}^{n_t-1} r_{t+i} + \gamma \max_{a'} \mathrm{Q}(S_{t+n_t}, a'; W^1, W^2) - \mathrm{Q}(S_t, a_t; W^1, W^2))^2] \tag{2}$$

$$\begin{aligned}\Delta(W^1, W^2) =& \alpha[\sum_{i=0}^{n_t-1} r_{t+i} + \gamma \max_{a'} \mathrm{Q}(S_{t+n_t}, a'; W^1, W^2) \\ & - \mathrm{Q}(S_t, a_t; W^1, W^2)]\nabla_{(W^1, W^2)} \mathrm{Q}(S_1, a_t; W^1, W^2)\end{aligned} \tag{3}$$

In Eq. (2), γ is the discount factor used to balance the weights of immediate rewards and future rewards, and E represents the expectation function. L evaluates the expected loss between the estimated reward $\mathrm{Q}(S_t, a_t)$ returned by DQN and the approximate reward $\tilde{\mathrm{Q}}$ (S_t, a_t) estimated by n-step Q-learning, as well as the estimated reward $\mathrm{Q}(S_{t+n}, a_{t+n})$ in the batch sample B. In Eq. (3),α represents the learning rate, which controls the step size of each update. $\nabla_{(W^1, W^2)}$ represents the deviation with respect to (W^1, W^2).

Exploration Module. In PER-GDRL, we designed an exploration module aimed at providing richer experience in updating the DQN parameters. In each training, we will first gradually add selected non-seed nodes to an empty seed set, and when each addition adds seed nodes to the seed set, we can get a $(S_t, a_t, r_t, S_{t+1}, I_{t+1})$ sample and push it into the replay buffer.

Random Walk Strategy. For the random walk strategy, each non-seed node added to the seed set is chosen completely at random. This strategy can greatly enrich the replay buffer, however, this can lead to some issues. If the nodes added to the seed set have low individual influence, it may result in generally low-quality samples being added to the replay buffer.

Preference Walk Strategy. The preference walk strategy calculates the influence of each node and determines the probability of each node being selected into the seed set. For a node v_j, the probability C_j of being selected is calculated as $C_j = (\widehat{\sigma}(v_j) - 1)/\sum_{i=1,2,\ldots,n}(\widehat{\sigma}(v_i) - 1)$. For each exploration,$K$ nodes are selected without replacement into the seed set according to their selection probabilities C. The selected nodes are then added to the seed set one by one in K steps. However, if we only use the preference walk strategy, it will tend to select nodes with larger degree into the seed set, and the experiences in the replay buffer will tend to be similar as the number of iterations increases. Therefore, we need to combine the two exploration strategies.

Genetic Operations. The genetic algorithm searches for the optimal solution by simulating the natural evolution processes of selection, crossover, and mutation. In this study, we use the genetic algorithm to explore the weights and bias parameters of the DQN neural network.

Crossover Operator. In the proposed algorithm, the crossover operation is applied to the weights and bias parameters of the DQN neural network. Specifically, two individuals are first selected, referred to here as $g_i(q)$ and $g_j(q)$, representing two DQN networks. For the two selected DQN networks, with a probability P_c, two equal-sized segments of their weights or bias parameters are randomly selected and exchanged, resulting in two new DQN networks $g_{i'}(q)$ and $g_{j'}(q)$.

Mutation Operator. In this study, we set a probability P_m to determine the likelihood of each gene undergoing mutation. For a selected individual $g_i(q)$, with a probability P_m, its weights or bias parameters are replaced with random numbers following a Gaussian distribution, resulting in a new DQN network $g_{i'}(q)$.

3.2 The Framework

In summary, the discrete seed determination problem is transferred into a continuous DQN parameter optimization problem. A population of DQN parameters is obtained, and the algorithm iteratively optimizes the DQN parameters of each individual through a combination of genetic algorithms and deep reinforcement learning strategies. Ultimately, the optimal DQN parameters are obtained, and the most influential seed set is selected through the Markov seed selection process. The algorithm flowchart is shown in Fig. 2. And the framework of the PER-GDRL algorithm is shown in Algorithm 1.

Algorithm 1 PER- GDRL

Input:
The input network: G_{in}; Size of the seed set: K; Number of iterations: $MaxGen$;
Size of population: n_g; Crossover probability: P_c; Mutation probability: P_m;

Output:
Best seed set: S; The influence of optimal seed set: $\hat{\sigma}(S)$;

1. Initialize a population $G_0(q)$ of size n_g with random DQN weight parameters; $i = 1$.
2. If $i \leq \frac{MaxGen}{2}$, employ the preference walk strategy; Else, employ the random walk strategy.
3. Obtain a new population $G_n(q)$ by performing crossover and mutation on $G_c(q)$ based on probabilities p_c and p_m.
4. Select the top $\frac{n_g}{2}$ individuals from both $G_c(q)$ and $G_n(q)$ to enter $G_c'(q)$, and randomly choose another $\frac{n_g}{2}$ individuals from the unselected individuals in $G_c(q)$ and $G_n(q)$ to enter $G_c'(q)$.
5. Select the best individual from $G_c'(q)$ and update its DQN parameters according to Eq. (12)(13)(14) and randomly replace the individuals that entered $G_c'(q)$ in the second step of Step 4 with the updated individuals; $i = i + 1$; $G_c(q+1) = G_c'(q)$.
6. If $i \geq MaxGen$, output the optimal seed set S and the 2-hop influence of optimal seed set $\hat{\sigma}(S)$; Else, turn to Step 2.

4 Experiments

In this section, experiments are shown to verify the effectiveness of PER-GDRL on synthetic and real networks of different scales. Three different synthetic networks with 200 or 500 nodes and a real traffic network with 248 nodes are considered. For each network, PER-GDRL is implemented and compared with the baseline method SDNE-GDRL [10], Degree selection method and PageRank [2]. Additionally, the performance of different seed set sizes has also been explored to demonstrate the superiority of the proposed algorithm.

In addition to this, we also compare the effect of PER-GDRL and the original algorithm without adding the PER method ranging under different coding methods. We apply four encoding methods (SDNE, LINE, node2vec, DeepWalk) to the original GDRL algorithm and PER-GDRL, in order to demonstrate the significance of the proposed algorithm incorporating the PER method as well as its independence from different encoding methods.

4.1 Experimental Settings

Network Data. In this experiment, we generated BA networks of 200 and 500 nodes with $\gamma = 3$, WS network of 200 nodes with $k = 4$ and $p = 0.1$ and ER network of 200 nodes with the connection probability of 0.1 The real network GM is a real traffic network, it has a total of 248 nodes and 390 edges.

Experimental Criteria. To evaluate the performance of the algorithms in part B, the 2-hop influence calculation method defined in Eq. (1) has been employed. Five independent experiments are conducted on different networks and seed scales, and the best results are recorded. We obtained the final ranking of each model by averaging the rankings of each model in each experiment based on the sum of the rankings in Table 1.

Simulation Settings. For all the deep reinforcement learning methods, the deep reinforcement learning and genetic operations are conducted for a total of 100 rounds, with the population size set to 50. In each round, the exploration strategy iterates for 20 times, and 10 individuals are randomly selected from the population for reinforcement learning updates. The input network was represented as a 2-dimensional vector. For PageRank [2], its parameter α is set to 0.85.

4.2 Experimental Results

From the results in Table 1, it can be seen that in most cases, the performance of the seed sets chosen by PER-GDRL is superior to that of other algorithms. The average ranking of the algorithms are as follows: PER-GDRL is ranked 1, SDNE-GDRL is ranked 1.6, Degree is ranked 3.6 and PageRank is ranked 3.4. Based on the average ranking it can be concluded that our proposed algorithm is ranked first in the final ranking, which proves its competitive performance and robustness.

In Fig. 3, the iterative curves of the maximum influence obtained of different seed set sizes on the ER and BA (500) networks for SDNE-GDRL and PER-GDRL are

Table 1. Experimental Results of Four Models on BA, WS, ER, BA (500) and GM Networks, Bold Scores are the Best in Method Group

Seed Set Size	Network	PER-GDRL	SDNE-GDRL	Degree	PageRank
20	BA	**22.1390**	22.1289	22.0556	22.0973
	WS	**21.1136**	**21.1136**	20.9455	21.0279
	ER	**21.5040**	21.4835	21.3992	21.3977
	BA500	**23.7118**	23.6454	23.6325	23.6310
	GM	**21.0229**	**21.0229**	20.8984	20.9173

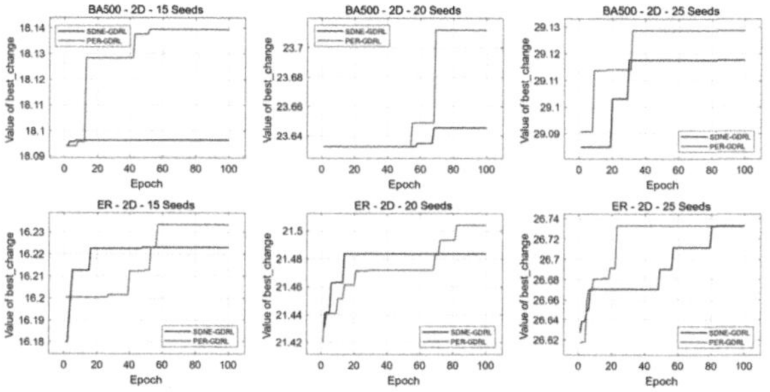

Fig. 3. Changes of SDNE-GDRL and PER-GDRL in the Maximum Influence of 15, 20 and 25 Seeds Selected on BA (500) and ER Network

presented. Since the optimal DQN parameters within the population are used after each iteration to calculate the maximum influence, the maximum influence obtained may remain unchanged for a considerable period. From Fig. 3, we can see that the proposed method has a significant advantage on the BA (500) and ER networks. The reason is that the BA (500) network is with a relatively large scale, providing numerous alternatives to select seeds, and the favorable of PER-GDRL is demonstrated. The ER network, compared to other networks, has more isolated nodes, and PER-GDRL also has an advantage in selecting seeds by bypassing these isolated nodes.

In Table 2, we apply different encoding methods to the proposed method and the original GDRL method without adding the PER method, which we call GDRL in Table 2. The algorithm proposed in this paper outperforms the original GDRL method without incorporating the PER method in most of the cases when different encoding methods are used. This proves the advantage of PER-GDRL in deep reinforcement learning and Markov seed selection module by applying PER method improvement, and it also proves that the superiority of PER-GDRL is not limited to a specific encoding method, but has wide applicability.

In conclusion, PER-GDRL has superiority over the other three existing approaches in solving the influence maximization problem. Moreover, the performance of PER-GDRL significantly surpasses that of other algorithms on large-scale networks and those with isolated nodes, which aligns with the current trend of increasing network complexity in

Table 2. Experimental Results of Two Models on BA, WS, ER, Bold Scores are the Better Scores for Two Algorithms and Underlined Scores are the Best Scores in the Method Group

Seed Set Size	Network	Encoding method	PER-GDRL	GDRL
10	BA	SDNE	**11.7230**	**11.7230**
		LINE	11.7221	**11.7222**
		Node2vec	**11.7222**	11.7105
		DeepWalk	**11.7222**	11.7105
	WS	SDNE	**10.5941**	**10.5941**
		LINE	**10.5939**	10.5938
		Node2vec	**10.5940**	10.5932
		DeepWalk	**10.5939**	10.5938
	ER	SDNE	**10.9100**	10.9098
		LINE	**10.8991**	10.8983
		Node2vec	**10.9099**	10.8994
		DeepWalk	**10.9096**	10.6096

practical applications. Also, the PER method is shown to be an explicit enhancement to the original GDRL and the proposed algorithm PER-GDRL is shown to be stable across different encoding methods.

5 Conclusion

In this paper, we propose PER-GDRL based on existing studies [2, 10]. By improving the reinforcement learning strategy, PER-GDRL achieves better performance and robustness in solving the influence maximization problem, which has been validated on synthetic and practical networks. An efficient solver is given to decision makers to tackle the information diffusion dilemma in practical applications.

In future work, more complicated scenarios where certain nodes may become unavailable due to attacks are considerable. How to select seeds in dynamical networks and enhance the robustness of seeds are worth of further investigations.

Acknowledgment. This work was supported in part by National Natural Science Foundation of China under Grant 62203477, in part by Guangdong Basic and Applied Basic Research Foundation under Grant 2024A1515010238, and in part by Tongchuang Intelligent Medical Interdisciplinary Talent Training Fund of Sun Yat-sen University.

References

1. Wang, S., Tan, X.: Solving the robust influence maximization problem on multi-layer networks via a Memetic algorithm. Appl. Soft Comput. **121**, 108750 (2022)

2. Page, L., Brin, S., Motwani, R.: The pagerank citation ranking: bringing order to the web. Technical report, Stanford InfoLab (1998)
3. Zhang, L., Yutong, L., Cheng, F.: A local-global influence indicator based constrained evolutionary algorithm for budgeted influence maximization in social networks. IEEE Trans. Netw. Sci. Eng. **8**(2), 1557–1570 (2021)
4. Zhang, Q.W., Bai, Q.H.: A discrete particle swarm optimization algorithm based on neighbor cognition to solve the problem of social influence maximization. J. Comput. **33**(4), 107–119 (2022)
5. Wang, S., Liu, J., Jin, Y.: Finding influential nodes in multiplex networks using a memetic algorithm. IEEE Trans. Cybernet. **51**(2), 900–912 (2019)
6. Yang, S., Du, Q., Zhu, G.: Balanced influence maximization in social networks based on deep reinforcement learning. Neural Netw. **169**, 334–351 (2024)
7. Lee, J.R., Chung, C.W.: A fast approximation for influence maximization in large social networks. In: Proceedings of the 23rd International Conference on World Wide Web, pp. 1157–1162 (2014)
8. Zhang, L., Liu, Y., Yang, H.: Overlapping community-based particle swarm optimization algorithm for influence maximization in social networks. AAI Trans. Intell. Technol. **8**(3), 893–913 (2023)
9. Li, H., Xu, M., Bhowmick, S.: PIANO: influence maximization meets deep reinforcement learning. IEEE Trans. Comput. Soc. Syst. **10**(3), 1288–1300 (2023)
10. Qumu, S., Zhu, C., Luo, Q., Wang, S.: Solving the influence maximization problem using a genetic deep reinforcement learning approach. In: 2023 5th International Conference on Data-driven Optimization of Complex Systems (DOCS), pp. 1–8. IEEE (2023)
11. Wang, D., Cui, P., Zhu, W.: Structural deep network embedding. In: Proceedings of the 22nd ACM SIGKDD International Conference on Knowledge Discovery and Data Mining, pp. 1225–1234 (2016)
12. Wang, S., Liu, J., Jin, Y.: Surrogate-assisted robust optimization of large-scale networks based on graph embedding. IEEE Trans. Evol. Comput. **24**(4), 735–749 (2019)

Continual Knowledge Graph Embedding by Structural Feature Distillation

Chong Mu[1,3], Zhao Kang[2,3], Qianghua Yuan[2], Sicong Liu[2], and Lizong Zhang[2,3](✉)

[1] School of Information and Software Engineering, University of Electronic Science and Technology of China, Chengdu 610000, Sichuan, China

[2] School of Computer Science and Engineering, University of Electronic Science and Technology of China, Chengdu 610000, Sichuan, China

l.zhang@uestc.edu.cn

[3] KASH Institute of Electronics and Information Industry, Kashi 844000, Xinjiang, China

Abstract. This paper presents a novel Continual Knowledge Graph Embedding (CKGE) approach that integrates local structural feature distillation to address the limitations of traditional static KGE methods. Our model dynamically updates embeddings without full retraining, reducing computational costs while retaining previously learned information. By employing a weighted sampling strategy based on betweenness centrality and a knowledge distillation mechanism that emphasizes entity-local structure similarity, our CKGE method effectively mitigates catastrophic forgetting. Experiments on four benchmark datasets reveal that our model surpasses existing baselines, showcasing its efficiency and effectiveness in continually learning knowledge graphs.

Keywords: Knowledge Graph Embedding · Continual Learning · Distillation

1 Introduction

Knowledge Graph Embedding (KGE) [1] maps entities and relations into a vector space, essential for tasks like link prediction and question-answering. However, traditional methods are static and require retraining with the addition of new knowledge, which is computationally expensive. Continual Knowledge Graph Embedding [2] (CKGE) has become an essential strategy. The objective of CKGE [3] is to integrate new knowledge into the model while retaining previously learned information, thus reducing the computational and temporal costs associated with retraining processes.

Regularization-based methods [2–6] within CKGE, such as the LKGE [3] model, use encoders and transfer strategies to maintain structural information while acquiring new knowledge. The IncDE model [4] further enhances this by using hierarchical ordering to determine the learning sequence for new triples, crucial for assimilating new information effectively. However, existing Regularization-based methods have primarily considered the semantic features of entities or relations, neglecting the explicit transfer of the topological structure features learned by the model.

G. Jeon and X. Kong (Eds.): ACIIBD 2024, LNICST 616, pp. 98–104, 2025.
https://doi.org/10.1007/978-3-031-91853-7_11

Therefore, this paper proposes a novel CKGE approach that incorporates the distillation of local structural features, enhancing the model's retention of learned structural features, particularly important nodes within the topology. Our model orders and learns embeddings according to the order in which the triples are added. A weighted sampling strategy prioritizes old triples with new entities, using betweenness centrality for weighting. Selected triples are then subjected to a knowledge distillation mechanism that leverages the similarity between entities and their local structural features to impose regularization constraints to preserve existing knowledge. Experimental results on four benchmark datasets show that our model outperforms all baselines.

In summary, the contributions of this paper are summarized as follows:

- We propose a novel CKGE method that distills local structural features, enhancing the ability to retain structural characteristics within evolving knowledge graphs.
- A weighted sampling strategy is introduced, prioritizing triples from the existing base that connect to entities in the new snapshot. We implement a distillation mechanism that leverages the similarity between entities and their local structural features. It effectively preserves old knowledge.
- Comprehensive experiments on four benchmark datasets show that our model outperforms existing baseline models.

2 Related Work

CKGE methods have been categorized into three distinct approaches: architecture-based, replay-based, and regularization-based methods. Architecture-based methods [7] adjust the neural network architecture to integrate new knowledge without forgetting previous tasks. Replay-based methods [8, 9] re-engage old knowledge during the training of new snapshot to prevent catastrophic forgetting. Regularization-based methods [2–6, 10, 11] stabilize learning by constraining weight updates or employing knowledge distillation, guiding the model to maintain essential parameters while adapting to new data. Several approaches [6, 12] integrate the ideas of regularization and replay to ensure the retention of historical knowledge while also incorporating regularization techniques to constrain modifications to critical model parameters.

3 Preliminaries

Growing Knowledge Graph. A KG is represented as $G = (E, R, T)$. Each triple in T is a composition of a head entity $h \in E$, a relation $r \in R$, and a tail entity $t \in E$, such that $(h, r, t) \in T$. As the KG evolves over time, reflecting the addition of new knowledge, this growth is denoted by $G_i = (E_i, R_i, T_i)$ at time i. The incremental changes are represented by $\Delta T_i = T_i - T_{i-1}$ for triples, $\Delta E_i = E_i - E_{i-1}$ for entities, and $\Delta R_i = R_i - R_{i-1}$ for relations.

Continual Knowledge Graph Embedding. Given a knowledge graph G, the objective of KGE is to project the triple into a lower-dimensional vector space R. The objective of CKGE is to seamlessly integrate new knowledge, denoted by the incremental sets ΔE_i, ΔR_i and ΔT_i, while retaining the learned representations of existing entities and relations from previous time steps, represented by E_{i-1} and R_{i-1}.

4 Methodology

The framework of the model is depicted in Fig. 1. The following sections will elaborate on each of these components in detail.

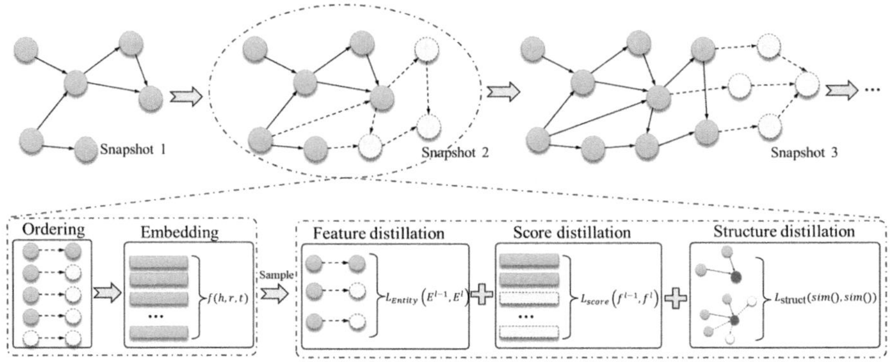

Fig. 1. Overview of the proposed model.

4.1 Hierarchical Ordering Module

In this section, we use a hierarchical ordering strategy as with the IncDE model. We initiate the layering by applying the Breadth-First Search (BFS) algorithm to the set of triples ΔT_i. We commence with the existing graph denoted as l_0. Subsequently, we identify all triples from ΔT_i that contain entities present in the old graph l_0 and categorize them as the next layer l_1. In next iterations, we treat the newly added entities from l_1 as old entities and repeat the above process. This iterative approach continues until all triples are appropriately categorized into layers.

Within each layer, we calculate the betweenness or node centrality which is used to measured relative importance of a node or edge. The node centrality measures the relative importance of a node. It is defined as $C_n(e) = \frac{f_{nei}(e)}{N-1}$. The $f_{nei}(e)$ is the number of neighbors of node e. N is the number of triples in E_i. The betweenness centrality $C_B(e)$ for a node e or an edge r is given by $C_B(e) = \sum_{h,t \in V, h \neq t} \frac{\sigma(h,t|e)}{\sigma(h,t)}$. The $\sigma(h, t|e)$ is the number of those shortest paths that pass through the node e or edge r.

The final importance score $a(h, r, t)$ of each triple can be computed as:

$$a(h, r, t) = max(C_n(h) + C_B(h), C_n(t) + C_B(t)) + C_B(r) \quad (1)$$

The importance score $a(h, r, t)$ aids in the prioritization of triples within each layer, ensuring that the more important triples are given prominence in the learning process.

4.2 Weighted Sampling and Local Structural Distillation Module

To preserve the model's learned structural features from old knowledge, we employ a novel distillation mechanism for entities in the current layer that have appeared in previous layers. We first select a set of K old triples $|T_l^K|$ through the weighted sampling based on $a(h, r, t)$, which include at least one entity present in l layer. Overall, the distillation mechanism contains three distillation loss, namely entity feature distillation, score distillation, and local structural feature distillation. The distillation loss function is based on the Huber Loss, which is formulated as:

$$L_{HuberLoss}(y_1, y_2) = \begin{cases} \frac{1}{2}(y_1, y_2), |y_1, y_2| \leq 1 \\ (y_1, y_2) - \frac{1}{2}, |y_1, y_2| > 1 \end{cases}. \tag{2}$$

Entity Features Distillation Loss: For an entity $i \in T_l^K$. that has been encountered in previous layers, we perform a weighted distillation of its features. It is defined as:

$$L_{Entity} = \sum_{i}^{|T_l^K|} W_i \cdot \alpha_i \cdot L_{HuberLoss}\left(e_i^{l-1}, e_i^l\right). \tag{3}$$

where $\alpha_i = C_n(e_i) + C_B(e_i)$, and $W \in R^{1\times|E_i|}$ is a learnable matrix. If the entity i has not appeared in any layer before the current one, the α_i is 0.

Score Distillation Loss: The objective is to ensure that the model's scoring ability for old triples remains consistent across layers. The loss is defined as:

$$L_{Score} = \sum_{(h,r,t)}^{T_l^K} L_{HuberLoss}\left(f^{l-1}(h, r, t), f^l(h, r, t)\right). \tag{4}$$

If entity i is absent in layer $l - 1$, the feature from the current layer is substituted to mitigate the impact of catastrophic forgetting.

Local Structure Distillation Loss: The core of this loss lies in quantifying the consistency between an entity's feature and its local structure feature, thereby ensuring that the model does not forget existing structural features when absorbing new knowledge. The loss is defined as:

$$L_{\text{struct}} = L_{HuberLoss}\left(sim\left(e_i^{(l-1)}, s_i^{(l-1)}\right), sim\left(e_i^{(l)}, s_i^{(l)}\right)\right). \tag{5}$$

Here, $sim()$ is a cosine similarity function, measuring the closeness between the entity's embedding and its local structural features. Inspired by the scoring function of TransE, the local structural features $s_i^{(l)}$ for an entity e_i is defined as:

$$s_i^{(l)} = \frac{1}{|N_h^l(e_i)|} \sum\nolimits_{(h,r,t)\in N_h^l(e_i)} (h_i + r_i), \tag{6}$$

where $N_h^l(e_i)$ denotes the h hops neighborhood of entity e_i at $l - th$ layer.

This loss function is designed to minimize the change in similarity between an entity's embedding and its local structure across layers, thereby preserving the model's understanding of the entity's context within the knowledge graph as it evolves.

In this paper, we employ TransE as the foundational embedding module. The scoring function for TransE is defined as $f(h, r, t) = |\mathrm{h} + \mathrm{r} - \mathrm{t}|$. The loss function of the TransE is defined as follow:

$$L_{\text{emb}} = \sum\nolimits_{(h,r,t)\in\Delta T_i} \max(0, f(h, r, t) - f(h\prime, r, t\prime) + \gamma) \quad (7)$$

The overall distillation loss in our framework is a weighted sum of three distinct loss components, each tailored to preserve different aspects of knowledge within the model. This composite loss is defined as follow:

$$L = L_{\text{emb}} + \gamma(L_{Entity} + L_{Score} + L_{\text{struct}}) \quad (8)$$

The coefficients γ serves as weight that balance the contributions of the loss.

5 Experiments

5.1 Experiment Settings

Datasets. We conducted our experiments on the four benchmark datasets proposed in the LKGE [3] framework, namely ENTITY, RELATION, FACT, and HYBRID. Each dataset focuses on the incremental increase of a specific element: entities, relations, triples, and a hybrid scenario, respectively. The details are presented in Table 1.

Table 1. Statistical details of four growing KGs.

Dataset	G_1			G_2			G_3			G_4			G_5		
	$\|\Delta T_1\|$	$\|E_1\|$	$\|R_1\|$	$\|\Delta T_2\|$	$\|E_2\|$	$\|R_2\|$	$\|\Delta T_3\|$	$\|E_3\|$	$\|R_3\|$	$\|\Delta T_4\|$	$\|E_4\|$	$\|R_4\|$	$\|\Delta T_5\|$	$\|E_5\|$	$\|R_5\|$
ENTITY	46388	2909	233	72111	5817	236	73785	8275	236	70506	11633	237	47326	14541	237
RELATION	98819	11560	48	93535	13343	96	66136	13754	143	30032	14387	190	21,594	14541	237
FACT	62024	10513	237	62023	12779	237	62023	13586	237	62023	13894	237	62023	14541	237
HYBRID	57561	8628	86	20873	10040	102	88017	12779	151	103339	14393	209	40326	14541	237

Baselines and Settings. Our evaluation leveraged both Fine-tuning [3] and continual learning baselines. The continual learning models include dynamic architecture (PNN [7]), replay (EMR [8], DiCGRL [9]), and regularization (SI [10], EWC [11], LKGE [3], IncDE [4]) methods. We evaluated model performance on link prediction using MRR and Hit@1/10. The final metric is an average from time $[1, i]$. The γ is set to 0.01. All other settings were consistent with IncDE [4].

5.2 Results

Main Results. As demonstrated in Table 2, our model secures the top performance, significantly outperforming the Fine-tune model and all other CKGE baselines. There is a significant improvement, especially when compared to the Fine-tune model, highlighting the substantial enhancement in learning efficiency. A particular improvement is observed against the SOTA model IncDE [4], with our model showing consistent enhancements across diverse datasets. The most notable increase is in the Hits@1 metric, with a 9.9% improvement on the ENTITY and RELATION datasets. These results collectively validate our model's capability to preserve existing knowledge while assimilating new data.

Table 2. Comparative results of link prediction. Results are averaged over five runs.

Models	ENTITY			RELATION			FACT			HYBRID		
	MRR	Hits@1	Hits@10	MRR	Hits@1	Hits@10	MRR	Hits@1	Hits@10	MRR	Hits@1	Hits@10
Fine-tune	0.165	0.085	0.321	0.093	0.039	0.195	0.172	0.090	0.339	0.135	0.069	0.262
PNN	0.229	0.130	0.425	0.167	0.096	0.305	0.157	0.084	0.290	0.185	0.101	0.349
EMR	0.171	0.090	0.330	0.111	0.052	0.225	0.171	0.090	0.337	0.141	0.073	0.267
GiCGRL	0.107	0.057	0.211	0.133	0.079	0.241	0.162	0.084	0.320	0.149	0.083	0.277
SI	0.154	0.072	0.311	0.113	0.055	0.224	0.172	0.088	0.343	0.111	0.049	0.229
EWC	0.229	0.130	0.423	0.165	0.093	0.306	0.201	0.113	0.382	0.186	0.102	0.350
LKGE	0.234	0.136	0.425	0.192	0.106	0.366	0.210	0.122	0.387	0.207	0.121	0.379
IncDE	0.253	0.151	0.448	0.199	0.111	0.370	0.216	0.128	0.391	0.224	0.131	0.401
Ours	**0.270**	**0.166**	**0.468**	**0.214**	**0.122**	**0.394**	**0.219**	**0.129**	**0.395**	**0.231**	**0.141**	**0.411**

Ablation Studies. The ablation results in Table 3 provide critical insights into the contribution of different loss. The results reveal the distillation loss as essential for our model's superior performance. Removing any component leads to a decrease in performance, with the model without distillation suffering the most. The full model, including all distillation losses, consistently outperforms, confirming the necessity of each loss for effective continual learning in knowledge graphs.

Table 3. Results of ablation study about the distillation loss.

Models	ENTITY			RELATION			FACT			HYBRID		
	MRR	Hits@1	Hits@10	MRR	Hits@1	Hits@10	MRR	Hits@1	Hits@10	MRR	Hits@1	Hits@10
w/o distill	0.190	0.100	0.355	0.136	0.079	0.255	0.167	0.090	0.320	0.188	0.106	0.342
w/o L_{Entity}	0.253	0.153	0.446	0.173	0.095	0.325	0.200	0.116	0.368	0.214	0.128	0.382
w/o L_{Score}	0.263	0.162	0.458	0.210	0.116	0.385	0.211	0.121	0.389	0.222	0.131	0.402
w/o L_{struct}	0.267	0.161	0.460	0.211	0.120	0.389	0.214	0.124	0.390	0.228	0.138	0.403
Ours	**0.270**	**0.166**	**0.468**	**0.214**	**0.122**	**0.394**	**0.219**	**0.129**	**0.395**	**0.231**	**0.141**	**0.411**

Table 4's ablation study on sampling strategy demonstrates a performance improvement with the weighted approach over both using all triples and random sampling. The results suggest that accounting for topological significance in triple selection slightly enhances the model's ability to predict links accurately.

Table 4. Results of ablation study about the sampling strategy.

Models	ENTITY			RELATION			FACT			HYBRID		
	MRR	Hits@1	Hits@10	MRR	Hits@1	Hits@10	MRR	Hits@1	Hits@10	MRR	Hits@1	Hits@10
All	0.269	0.164	**0.468**	0.213	**0.122**	0.392	0.219	**0.130**	0.395	0.230	0.139	**0.413**
Random	0.268	0.163	0.465	0.211	0.120	0.391	0.218	0.127	0.393	0.229	0.137	0.408
Weighted	**0.270**	**0.166**	**0.468**	**0.214**	**0.122**	**0.394**	**0.219**	0.129	**0.395**	**0.231**	**0.141**	**0.411**

6 Conclusion

In conclusion, this paper introduces an innovative Continual Knowledge Graph Embedding approach that efficiently integrates new knowledge without retraining, significantly outperforming existing methods on benchmark datasets. By leveraging local structural feature distillation and a weighted sampling strategy, our model preserves old knowledge while accommodating new knowledge.

References

1. Bordes, A., Usunier, N., Garcia-Duran, A., Weston, J., Yakhnenko, O.: Translating embeddings for modeling multi-relational data. Adv. Neural Inf. Process. Syst. **26** (2013)
2. Daruna, A., Gupta, M., Sridharan, M., Chernova, S.: Continual learning of knowledge graph embeddings. IEEE Rob. Autom. Lett. **6**, 1128–1135 (2021)
3. Cui, Y., et al.: Lifelong embedding learning and transfer for growing knowledge graphs (2022)
4. Liu, J., et al.: Towards continual knowledge graph embedding via incremental distillation. In: Proceedings of the AAAI Conference on Artificial Intelligence, vol. 38, no. 8, pp. 8759–8768 (2024)
5. Liu, J., Wang, P., Shang, Z., Wu, C.: IterDE: an iterative knowledge distillation framework for knowledge graph embeddings (2023)
6. Wei, D., Gu, Y., Song, Y., Song, Z., Li, F., Yu, G.: IncreGNN: incremental graph neural network learning by considering node and parameter importance. In: Database Systems for Advanced Applications, Lecture Notes in Computer Science, pp. 739–746 (2022)
7. Rusu, A.A., et al.: Progressive neural networks. arXiv preprint arXiv:1606.04671 (2016)
8. Wang, H., Xiong, W., Yu, M., Guo, X., Chang, S., Wang, W.Y.: Sentence embedding alignment for lifelong relation extraction. In: Proceedings of the 2019 Conference of the North. Presented at the Proceedings of the 2019 Conference of the North, Minneapolis, Minnesota (2019)
9. Kou, X., Lin, Y., Liu, S., Li, P., Zhou, J., Zhang, Y.: Disentangle-based continual graph representation learning. In: Proceedings of the 2020 Conference on Empirical Methods in Natural Language Processing (EMNLP) (2020)
10. Zenke, F., Poole, B., Ganguli, S.: Continual learning through synaptic intelligence. In: International Conference on Machine Learning, pp. 3987–3995. PMLR (2017
11. Kirkpatrick, J., et al.: Overcoming catastrophic forgetting in neural networks. Proc. Natl. Acad. Sci. **114**, 3521–3526 (2017)
12. Hoang, T.D., et al.: Universal graph continual learning. arXiv preprint arXiv:2308.13982 (2023)

TZRT-RPC: A Hybrid Multi-OS Real-Time Communication Scheme Based on TrustZone

Jintao Qiu, Pei Zhang(✉), Ji Zhang, and Long Zuo

Department of Basic Software, East China Institute of Computing Technology, Shanghai 201808, China
zpcorba@126.com

Abstract. With the rapid development of IoT, intelligence, functional safety and information security, a single operating system can no longer meet the diverse application requirements. In response to this challenge, the hybrid deployment of multiple operating systems on the same SoC platform has become a hot research direction. However, the current communication framework between multiple operating systems on the same hardware platform has limitations in concurrency and real-time performance, which limits the effective cooperation between different systems. Therefore, this study proposes an efficient hybrid multi-system communication framework based on TrustZone technology, called TZRT-RPC, which aims to optimize the real-time performance and concurrency of Remote Procedure Calls (RPCs) between different systems. By using SGI (Software Generated Interrupt) and Interrupt Handler RPC Servicing mechanism, the framework precisely controls the execution time of RPC service routines. Furthermore, by combining Scheduler Bypass and Dual Priority Ceiling strategy, TZRT-RPC reasonably distributes RPC requests to each core to achieve efficient concurrent processing of RPC requests. Experimental results show that TZRT-RPC can support high concurrent communication while ensuring real-time communication between systems.

Keywords: RPC · Real-Time · TrustZone · Communication

1 Introduction

In the field of automotive and avionics, a significant trend in the development of operating systems is the integration of various functions such as real-time control, visual operation and audio-visual entertainment on a single hardware platform. It has become a hot research direction to deploy a variety of different types of systems on the same SoC platform to form a hybrid multi-operating system [1]. Specifically, in automotive electronic systems, a typical architecture contains control software responsible for controlling electronic control units (ECUs), and extended software to handle entertainment information devices such as audio, video, and navigation (AVN) systems. The current technological evolution in the automotive market is moving towards closer integration of extension software with control software [2].

G. Jeon and X. Kong (Eds.): ACIIBD 2024, LNICST 616, pp. 105–117, 2025.
https://doi.org/10.1007/978-3-031-91853-7_12

In the field of avionics, the ARINC 653 standard promotes the integration of flight control system, environmental control system and entertainment system on the same SoC platform through the application of virtualization technology [3]. It improves the efficiency of resource utilization and enhances the flexibility and scalability of the system.

However, virtualization technology introduces additional overhead. Full virtualization has the best compatibility, at the cost of significant I/O performance degradation [4]; Although there are some technologies to reduce the performance loss, such as para-virtualization and hardware-assisted virtualization, these technologies require the virtual machine to make a lot of changes to its own design, which has a negative impact on compatibility.

To this end, a method of deploying subsystems in an isolated environment provided by hardware becomes a possible solution to the above problems. For example, the TZDKS model [11] uses TrustZone technology to divide system resources into secure domain and non-secure domain to ensure that RTOS performs critical tasks in the secure domain. In contrast, the more fully featured GPOS are deployed in the non-secure domain and participate in the scheduling as idle tasks of the RTOS. In such a hybrid Multi-OS architecture, real-time communication between systems is the basis of efficient collaboration. For example, in the scenario of autonomous driving system, the collision detection task in RTOS not only needs to collect and analyze sensor data, but also relies on the processing of short-term path planning data. Typically, path planning tasks are supported by GPOS, which have more resources. In this process, the interaction of GPOS providing services to RTOS inevitably relies on the support of Remote Procedure Call (RPC) mechanism.

This paper presents a hybrid multi-OS real-time communication scheme implemented in TZDKS, which uses Software Generated Interrupt and interrupt handling routine RPC mechanism to ensure the real-time performance of RPC process, and makes full use of multi-core advantages to realize good support for concurrent RPC requests.

2 Background

2.1 TrustZone Technology

TrustZone [6] is a system security extension technology developed by ARM company. The main goal is to ensure the security of the embedded system and prevent the leakage of security-sensitive data or the attack of malicious programs on key functions in the system. TrustZone virtualizes the same CPU into two isolated environments: a secure world and a normal world. The resources in the secure world are protected by hardware, and the normal world is strictly forbidden to access the protected resources in the secure world.

2.2 TZDKS Architecture

TZDKS [11] is an architecture that integrates RTOS and GPOS efficiently by using the isolation mechanism and security interrupt mechanism in TrustZone technology. Under the framework of TZDKS, RTOS runs in Secure World and is driven by an independent security clock. GPOS (e.g., Linux) runs in the Normal World and is driven by a normal clock. In RTOS, GPOS is regarded as Idle task to participate in scheduling. On the one hand, the independent security clock and priority-based preemptive scheduling of RTOS can ensure the real-time performance of all tasks in RTOS. On the other hand, GPOS can make full use of the idle time of RTOS to perform non-real-time tasks.

The working principle of TZDKS is shown in Fig. 1, when the CPU is in Secure World, if all the real-time tasks have been executed, the RTOS initiates the SMC request through the idle task to actively cede the usage of the CPU to GPOS. When the CPU is in the Normal World, the GPOS kernel can use the TrustZone driver to initiate the SMC request to return the CPU usage right to the RTOS. In addition, when the security clock interrupt arrives, the Monitor of EL3 will actively interrupt the execution of Linux, switch the CPU to Secure World, and hand over the security clock interrupt to the RTOS.

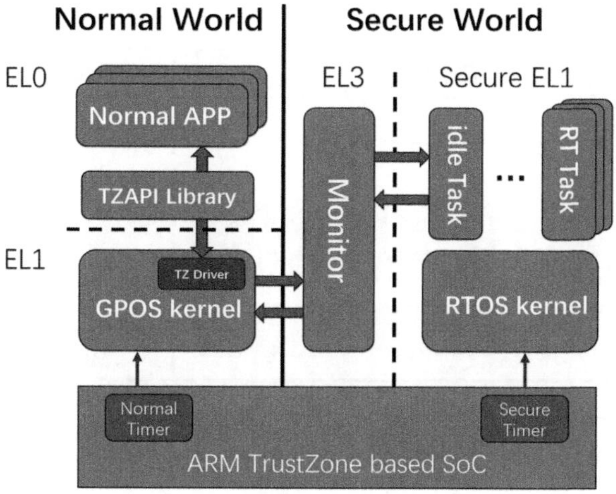

Fig. 1. TZDKS Architecture

2.3 Inter-OS Communication

In current hybrid multi-OS architectures, communication between different operating systems is usually carried out under the virtualization framework. In some virtualization solutions such as: Xen [5] and Jailhouse [8] have implemented the communication mechanism between different partitioned operating systems, but they are not suitable for the communication between dual systems under the TrustZone framework. In addition, there are some solutions that use related hardware mechanisms to achieve efficient communication, such as safeG [7] and TZ-virtio [9], which use SGI as the event path and shared memory as the data path. The above schemes only realize the availability of communication functions, and do not adopt targeted design for real-time communication. In the research of Pan Dong's team, a time predictability RPC scheme combining SGI and Interrupt Handler RPC Service Routine was proposed, namely RTRG-RPC [10]. This scheme can meet the real-time requirement of communication. However, when multiple real-time tasks initiate RPC requests concurrently, this scheme fails to effectively utilize the parallel processing ability of multi-core SoC.

3 Design of TZRT-RPC

3.1 Communication Mechanism

The initiator of RPC request in hybrid dual operating system can be either RTOS or GPOS, and when RTOS is the initiator, it has higher requirements for the time predictability of RPC request. TZRT-RPC is mainly designed and implemented for this type of request. Figure 2 shows the communication mechanism of TZRT-RPC, where by default the RTOS runs in Secure World on one of the cores of a multi-core CPU.

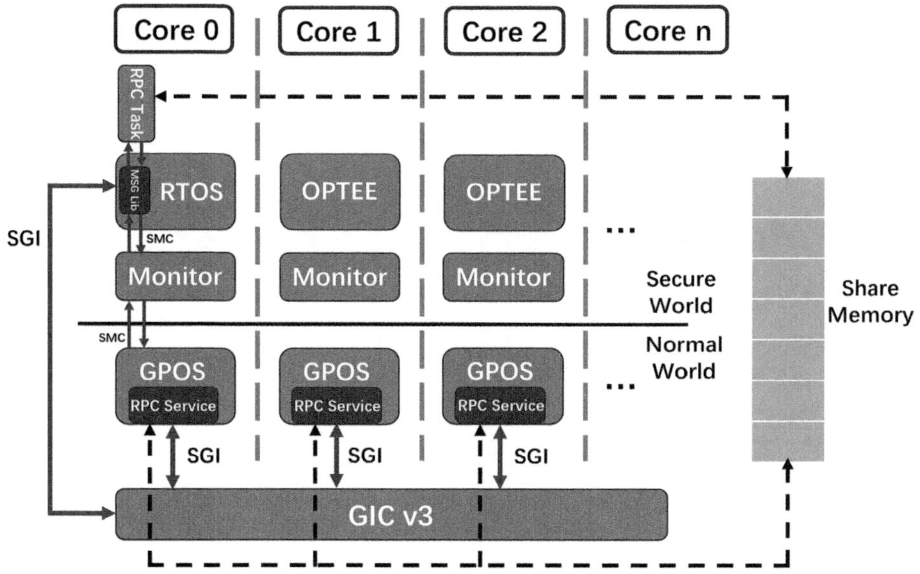

Fig. 2. TZRT-RPC Architecture

Data Path
TZRT-RPC is implemented based on shared memory. The shared memory area is statically configured at compile time, and this part of memory is configured as non-safe area memory by TZC during system boot, so that GPOS and RTOS can access it at the same time. After system startup, GPOS and RTOS establish mmu mapping of this memory area respectively.

Event Path
On the event path, TZRT-RPC uses the Software Generated Interrupt between Secure

World and Normal World to communicate with each other[1]. The injection of non-Secure Group 1 SGI signals processed by Normal World in Secure World can be achieved by configuring GICR_NSACR, and vice versa. When Normal World receives an SGI signal from Secure World, it executes the corresponding RPC service item according to the contents in the shared memory and notifies the RTOS through SGI after the service is completed.

In RTOS, TZRT-RPC allocates suitable cores for real-time tasks to execute RPC services when they make RPC requests. According to the allocation of CPU cores, it needs to be divided into two cases:

- Asynchronous Request: If the execution core of the RPC service is not the RTOS core, since the RPC service is asynchronous with the real-time task, the RPC request initiator only has to suspend itself after injecting the SGI and wake up when it receives the SGI signal indicating the completion of the RPC service.
- Synchronous Request: If the SGI signal sent to the Normal World is processed by the core where the RTOS is located, the RPC service and the real-time task are synchronized at this time, so the CPU needs to be switched to the Normal World after sending the SGI, so that GPOS can respond to the RPC request of the real-time task.

In the TZDKS model, when the CPU is in Secure World, the CPU only responds to the Secure Group1 type interrupt, while the SGI injected by the RPC request initiator (denoted as τ_{RPC}) is a non-Secure Group1 interrupt, which is masked by the current core. Therefore, after the interrupt injection, the CPU needs to be switched to Normal World by SMC instruction immediately, and the RPC service can be executed normally after the GPOS resumes execution. After GPOS completes the RPC service, it will also inject a Secure Group 1 type SGI sent to the current core through GIC v3. Since the CPU is in Normal World at this time, this interrupt will be captured by the Monitor of the current core and given to the RTOS in Secure World.

3.2 Real-Time Design

Through SGI and shared memory mechanism, it is possible to communicate between RTOS and GPOS. However, the existing design still has the problem of priority inversion, which prolongs the response time of RPC and reduces the resource utilization in some cases. TZRT-RPC uses a series of measures, such as interrupt service routine RPC and dual priority ceiling, to ensure the real-time performance of RPC during concurrency. The pseudocode in Fig. 3 presents the implementation of the above series of mechanisms.

Requirement Analysis and Concept Definition

To ensure that concurrent requests for RPC services do not affect the real-time performance of τ_{RPC}, the TZRT-RPC model must meet the following requirements:

1. Concurrency: Multiple tasks with different priorities can request the RPC service concurrently.

[1] IN TrustZone-enabled CPUs, Typically, interrupts 0–7 are configured as non-Secure Group 1 interrupts handled by GPOS, and interrupts 8–15 are configured as Secure Group 1 interrupts handled by TEE (Trusted Execution Environment) OS.

```
Result RPC_Request(void) {
    /* fault broadcast */
    if(check_fault(rpc_flag))
        return FAULT;
    /* allocate an available core */
    semaphore_take(&core_sem, portMAX_DELAY);
    cpu = get_available_core(rpc_flag);
    mutex_lock(rpc_mux[cpu], portMAX_DELAY);
    /* notify GPOS */
    raise_sgi(cpu, SGI_NUM);
    rpc_flag[cpu] = BUSY;
    if(cpu == current_cpu) {
        set_timeout(RPC_WCET);
        /* resume GPOS */
        while(rpc_flag[cpu] != FINISH && !is_timeout())
            switch_to_normal();
    }else {
        semaphore_take(&rpc_sem[cpu], RPC_WCET);
    }
    /* finish RPC */
    if(rpc_flag[cpu] != FINISH)
        rpc_flag[cpu] = FAULT; /* RPC Time Out */
    else
        semaphore_give(&core_sem);
    mutex_unlock(rpc_mux[cpu]);
    return rpc_flag[cpu];
}
```

Fig. 3. TZRT-RPC event path implementation

2. Time-predictable RPC: The worst-case execution time of the RPC service itself is predictable.
3. Real-Time: The worst-case time between the start of an RPC request and its completion is predictable.
4. Failure handling: The implementation requires a designated mechanism for handling failures.

In addition, for the convenience of subsequent explanations, some of the concepts involved in the following paragraphs are defined as follows:

1. RPC core: refers to the CPU core assigned to the task that initiates the RPC request, which is responsible for executing the corresponding RPC service. Each RPC core can execute any type of RPC service.
2. RPC task vs. Normal task: The task in an RTOS that need to initiate RPC requests are collectively referred to as RPC task, and the remaining tasks are collectively referred to as Normal tasks.
3. Synchronous RPC and Asynchronous RPC: For RPC tasks, when the obtained RPC core is the core where the RTOS resides, the acquired RPC core is referred to as

a Synchronous RPC core, the initiated RPC request is called a Synchronous RPC request, and the requested RPC service is termed Synchronous RPC service; conversely, they are respectively called Asynchronous RPC core, Asynchronous RPC request, and Asynchronous RPC service.

4. RPC service time and RPC response time: RPC service time refers to the time interval between the SGI signal initiated by the RPC task and the SGI signal received from the GPOS by the RTOS. This metric is used to evaluate the responsiveness of GPOS to RPC requests. RPC response time refers to the time elapsed from the time when an RPC task attempts to apply for an RPC core until the task releases the core, which represents the total duration of the RPC process, and is the key indicator to measure the real-time performance of TZRT-RPC model.

Scheduler Bypass Mechanism

In order to meet the multi-task concurrent request RPC service, TZRT-RPC no longer takes the idle task as the only container of GPOS. If τ_{RPC} initiates a synchronous RPC request, it will allocate its time slice to GPOS. And initiate an SMC request to switch CPU to Normal World immediately after injecting SGI signal. Before RPC is completed, τ_{RPC} will always participate in scheduling on behalf of GPOS.

Interrupt Handler RPC Servicing

Since GPOS is usually not a real-time system, the RPC service cannot exist as a process in GPOS. In TZRT-RPC, the interrupt service routine is used as the carrier of RPC service. Taking Linux as an example, the Interrupt Handling Process is divided into the upper half of hard interrupt handling and the lower half of soft IRQ handling. The hard interrupt handling routine is executed immediately when an interrupt arrives, while the more complex and time-consuming interrupt handling flow is placed in the soft IRQ handling routine and can be deferred.

By using the RPC service routine as a hardware interrupt handling routine in Linux, it is ensured that Linux will immediately begin executing the RPC service upon receiving the SGI. In addition, in order to ensure the timely response of GPOS to the SGI interrupt corresponding to RPC service, and the process of RPC service is not interrupted by other interrupts in GPOS, the interrupt corresponding to RPC service should be configured as the interrupt with the highest priority in non-Secure Group 1.

RPC Service Core Allocation

Compared with Synchronous RPC, Asynchronous RPC service does not occupy the time slice of RTOS. RTOS can use the saved time to schedule other tasks. Therefore, TZRT-RPC always gives priority to assigning available Asynchronous RPC cores to τ_{RPC}. Synchronous RPC cores are assigned to τ_{RPC} only if all asynchronous RPC cores are executing RPC services.

Dual Priority Ceiling

To avoid priority inversion, TZRT-RPC uses a priority ceiling policy. Once τ_{RPC} obtains the RPC core, it immediately sets its priority to the ceiling priority using the ceiling mutex. However, the event path of a Synchronous RPC request is not the same as that of an Asynchronous RPC request. This may result in the Asynchronous RPC request being blocked by the executing Synchronous RPC service after completion, delaying the release of the RPC core.

In this paper, we propose a dual priority ceiling strategy, which assumes that the following tasks exist in the system:

1. τ_{RPC}^{high}, which has the highest priority among the RPC tasks.
2. τ_{RPC}^{low}, which has the lowest priority among the RPC tasks.
3. τ_{Norm}^{high}, which has the highest priority among the Normal tasks.

The upper limit of system priority is p_{max}, the priority of task τ is denoted as $p(\tau)$, and the priority ceiling is denoted as $p_{ceiling}(\tau)$; Suppose that the above task satisfies the following constraints:

$$p_{max} > p\left(\tau_{RPC}^{high}\right) > p\left(\tau_{Norm}^{high}\right) \tag{1}$$

When the RPC core obtained by τ_{RPC} is a Synchronous RPC core, there is:

$$p_{ceiling}(\tau_{RPC}) = max\left(p\left(\tau_{Norm}^{high}\right) + 1, p(\tau_{RPC})\right) \tag{2}$$

If the obtained RPC core is an Asynchronous RPC core, there is:

$$p_{ceiling}(\tau_{RPC}) = max\left(p\left(\tau_{RPC}^{high}\right) + 1, p(\tau_{RPC})\right) \tag{3}$$

For Asynchronous RPC request, when the RPC service execution is completed, τ_{RPC} is the task with the highest priority in the system, which will immediately wake up and release the RPC core, and then restore the priority. If the priority of this task is lower than the ceiling priority in Eq. (2) and the current core is executing Synchronous RPC, τ_{RPC} will immediately preempt the task that made the Synchronous RPC request.

For Synchronous RPC requests, under the protection of the priority ceiling in Eq. (2), the RPC service process will not be interrupted by Normal task during the execution of the RPC service, so as to avoid the priority inversion phenomenon similar to that in the execution of Synchronous RPC service: the high-priority RPC tasks will be preemptive by Normal tasks with lower priority because they cannot apply for resources.

In addition, the lower priority ceiling of the Synchronous RPC tasks ensures that the Asynchronous RPC tasks can successfully release the RPC cores when the Asynchronous RPC service is completed. Ensuring that the RPC cores are available immediately after the RPC service is completed, which improves the resource utilization efficiency.

In Fig. 3, TZRT-RPC implements the dual priority ceiling mechanism using the ceiling mutex rpc_mux[cpu], and when an RPC task locks the RPC service through rpc_mux[cpu], the priority of the task is automatically promoted to the priority ceiling corresponding to the RPC core.

Fault Management Mechanism

As shown in Fig. 3, in multi-core CPU, TZRT-RPC uses rpc_flag to carry out unified fault management for all cores participating in RPC. Rpc_flag is an array with the length of the number of CPU cores, each item in the array represents the working state of the corresponding CPU core. The state transformation of each CPU core is shown in Fig. 4.

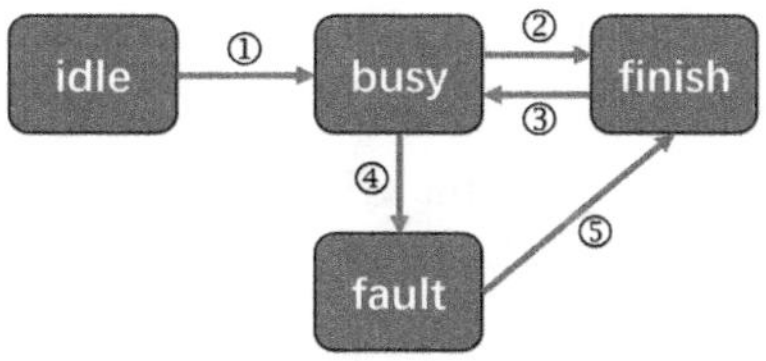

Fig. 4. RPC core state transition diagram

4. All cores are in the idle state before RPC service, and when a core is selected as an RPC core, it becomes busy.
5. The busy RPC core transitions to finish state when it completes RPC service on time.
6. When an RPC request is made, the CPU in finish mode can be selected as the RPC core and will be switched to busy mode when RPC service starts
7. If the RPC service exceeds the specified time limit, this RPC core is considered to be faulty and turns into fault state.
8. The RPC service completes successfully after timeout, and sends SGI signal to the RTOS, which considers that the core still has the service capability, so it turns to finish state.

In the fault management mechanism of TZRT-RPC, each RPC service has an upper limit of execution time, and the timing starts from the execution of RPC service. If the execution time of RPC service exceeds the upper limit, the RPC request will be considered to fail. Since GPOS provides additional assistance to the RTOS, failure does not have catastrophic consequences, and aborting failed RPCS allows other critical operations to continue on time.

When a core is marked as fault, it will not participate in the allocation of RPC cores until it is back to normal. Furthermore, if all cores fail[2], in the implementation of Fig. 3, τ_{RPC} will directly abandon the RPC request, further avoiding the impact of the failure on the real-time performance of the RTOS.

3.3 Analysis of RPC Service Time

Model Assumptions

In order to analyze the Worst-Case Response Time of the RPC service, we need to build a multi-task system model, which makes the following assumptions:

1. The interrupt handling overhead and scheduling overhead of RTOS are negligible.
2. There is no synchronization between RPC service routines. One RPC service will not be blocked by another RPC service.
3. The system is schedulable: as long as the RPC response time is controllable, all real-time tasks can be completed before their deadlines.

Where the assumption 1. Is related to the actual execution time of RPC. In this paper, Phytium FT-D2000 processor is used as an example to measure the actual execution time of the above two overheads respectively.

[2] E.g., GPOS completely crashes, in which case the failure will not propagate to the RTOS due to the isolation feature of TrustZone technology.

For the overhead of interrupt handling, the RTOS interrupt handling process needs to be divided into two cases:

1. Handling of Secure interrupts when the CPU is in Secure World
2. Handling of Secure interrupts when the CPU is in Normal World.

The latter needs to undergo world switching. The processing times of the two interrupts were recorded 100 times, respectively, and the results are shown in Table 1.

Table 1. Interrupt handling and scheduling overhead

	Maximum	Minimum	Average
Interrupt Handle from Secure World	0.481 us	0.126 us	0.228 us
Interrupt Handle from Normal World	3.875 us	1.062 us	1.556 us
Task Schedule Time	1.368 us	0.594 us	0.827 us

The experimental results show that the above two overheads are within 4us, which can be ignored compared with the processing delay in ms of the RPC service routine.

Worst-Case Response Time Analysis for RPC

For the convenience of calculation, it is defined as follows:

1. The RPC service time is denoted as $T_{Service}$. Thanks to the RPC mechanism of the interrupt service routine, $T_{Service}$ is only related to the RPC service itself and is a fixed value.
2. The RPC response time is denoted as T_{all}, which can reflect the actual response time of RPC service if the system is schedulable.
3. According to the dual priority ceiling mechanism, assuming that the RPC core assigned to τ_{RPC} is a Synchronous RPC core, the priority ceiling corresponding to τ_{RPC} is denoted as $p_{ceiling}^{low}(\tau_{RPC})$, otherwise it is denoted as $p_{ceiling}^{high}(\tau_{RPC})$.

Due to the limited number of cores in the system, when there are more RPC tasks, it is inevitable that τ_{RPC} will be blocked because it cannot obtain RPC cores. Therefore, the calculation of the worst response time of RPC should be discussed in two cases respectively:

1. When an RPC is initiated by τ_{RPC}, there are enough RPC cores available, and τ_{RPC} can immediately obtain available RPC cores:
   a. RPC core is Asynchronous RPC core. In this case, the priority of τ_{RPC} is promoted to $p_{ceiling}^{high}(\tau_{RPC})$, and it will go to sleep after SGI injection, and will be awakened on time when RPC is completed, in this case: $T_{all} = T_{Service}$.
   b. RPC core is Synchronous RPC core. According to the RPC core allocation policy, all cores are executing RPC service routines, and only task τ_{RPC}^{mid} with priority between $p_{ceiling}^{low}(\tau_{RPC})$ and $p_{ceiling}^{high}(\tau_{RPC})$ can interrupt the execution of τ_{RPC} and

be blocked because it cannot apply for RPC cores. Even if an Asynchronous RPC core is released during the execution of the Synchronous RPC service, τ_{RPC}^{mid} will still prefer the asynchronous RPC core. Ignoring the scheduling time, τ_{RPC}^{mid} cannot significantly affect τ_{RPC}, so: $T_{all} = T_{Service}$.

2. When τ_{RPC} initiates an RPC, the available RPC core is not immediately available and is therefore blocked until some other RPC task releases the RPC core. Under the dual priority ceiling mechanism, there must be: $p(\tau_{RPC}) > p\left(\tau_{Norm}^{high}\right) + 1$, and there may be more than one RPC task with priority higher than $p\left(\tau_{Norm}^{high}\right) + 1$. If at this time, m RPC tasks including τ_{RPC} are blocked because they cannot apply for RPC cores, and τ_{RPC} ranks the n.th among the m tasks according to their priority. When an RPC core is freed, the blocked tasks are woken up in order of priority and the freed RPC cores are allocated to them. Since the worst-case execution time of all RPC service is $max(T_{Service})$, the worst-case response time of an RPC request initiated by τ_{RPC} satisfies:

$$max(T_{all}) = \left(\left\lceil \frac{n}{A} \right\rceil + 1\right) \cdot max(T_{Service}) \tag{4}$$

Here, A is the number of cores in GPOS that can be used to perform RPC services, and n is necessarily no more than the number of RPC tasks with higher priority than $p\left(\tau_{Norm}^{high}\right) + 1$, which is also easy to determine.

In summary, regardless of whether it is blocked or not, as long as the system is schedulable, the RPC requests can be responded to in a predictable time.

4 Evaluation

We used the FT-D2000 processor produced by Phytium as the experimental platform. To measure the responsiveness of the RPC under high load, we used only two CPU cores including the current core for responding to RPC requests from the RTOS. And as shown in Table 2, we created four real-time tasks.

Among them, Task1, Task2 and Task4 are both RPC task, while Task3 is a Normal task, and its priority is only higher than Task4. Each RPC task will issue an RPC request in each execution cycle. The RPC Response Time T_{all} of RPC and the RPC Service Time $T_{Service}$ were recorded respectively. This four tasks were executed concurrently, and the recording stopped after all RPC tasks completed 100 RPC requests. The RPC result of the three tasks are shown in Fig. 5. The horizontal axis is the request sequence number. The vertical axis is the response time in nanoseconds.

By Eq. (4), it can be concluded that the worst response time of RPC is 20 ms. In the above experimental results, the service time of all RPCs is accurately controlled within 10.5 ms. Task1 and Task2 can keep most of the RPC response time the same as the RPC service time due to their higher priority. Due to its low priority, Task4 cannot request an RPC core most of the time and has to block until an RPC core becomes available. Even so, the vast majority of Task4 RPC requests were answered in a timely manner; only two RPC requests exceeded this time, which was very close to 20 ms and can be considered to be caused by scheduling overhead.

Table 2. Real-time Task information

Task Name	Task Parameters			
	Period	Priority	Execute time	$T_{Service}$
Task 1	25 ms	200	——	10 ms
Task 2	40 ms	150	——	10 ms
Task 3	20 ms	120	5ms	——
Task 4	50 ms	100	——	10 ms

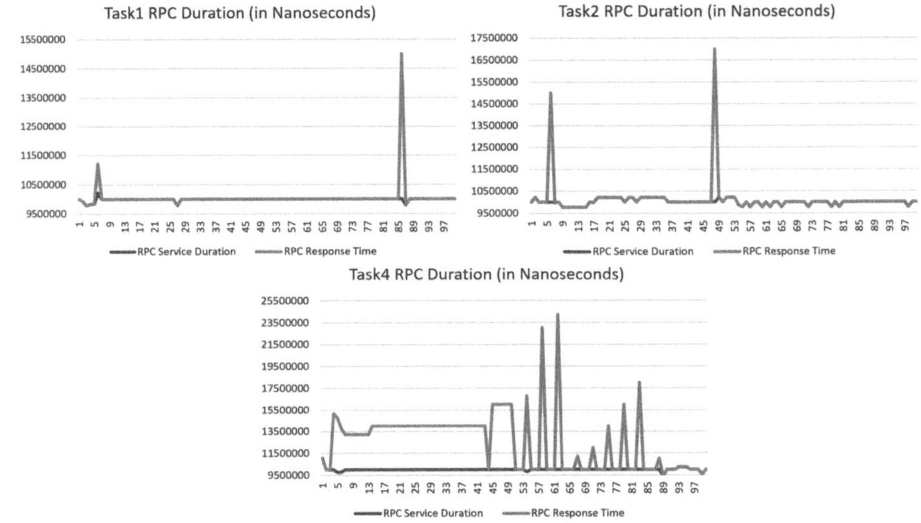

Fig. 5. RPC Response Time of Task1, Task2 and Task4

5 Conclusions

In this paper, we propose a hybrid multi-OS communication framework: TZRT-RPC, which can meet the real-time and high concurrency requirements simultaneously. Through the dual priority ceiling mechanism and a reasonable RPC core allocation scheme, the timely response of RPC requests in the case of high concurrency is realized, and the resource utilization is improved. Finally, by limiting system resources and simulating the response of RPC under high load, the effectiveness of TZRT-RPC framework is successfully proved.

References

1. Berger, R.: Consolidation in vehicle electronic architectures, Think: Aact (2015)
2. Jiang, Z., et al.: Bridging the pragmatic gaps for mixed-criticality systems in the automotive industry. IEEE Trans. Comput. Aided Des. Integr. Circuits Syst. **41**(4), 1116–1129 (2022). https://doi.org/10.1109/TCAD.2021.3075422

3. Chen, K., Du, C., Chen, J., Gu, Q.: Design of virtual simulation experiment platform based on ARINC 653 specification. In: 2020 IEEE 4th Information Technology, Networking, Electronic and Automation Control Conference (ITNEC), Chongqing, China, pp. 963–967 (2020). https://doi.org/10.1109/ITNEC48623.2020.9084836
4. Đorđević, B., Timčenko, V., Kraljević, N., Davidović, N.: File system performance comparison in full hardware virtualization with ESXi and Xen hypervisors. In: 2019 18th International Symposium INFOTEH-JAHORINA (INFOTEH), East Sarajevo, Bosnia and Herzegovina, pp. 1–5 (2019). https://doi.org/10.1109/INFOTEH.2019.8717664
5. Lesniak, F., Harbaum, T., Becker, J.: Low-latency inter-domain communication on the Xen hypervisor. In: 2023 IEEE 16th International Symposium on Embedded Multicore/Many-core Systems-on-Chip (MCSoC), Singapore, pp. 340–346 (2023). https://doi.org/10.1109/MCSoC60832.2023.00057
6. Wang, K.C.: ARM TrustZone and secure operating systems. In: Embedded and Real-Time Operating Systems. Springer, Cham (2023).https://doi.org/10.1007/978-3-031-287015_12
7. Sangorrin, D., Honda, S., Takada, H.: Reliable and efficient dual-OS communications for real-time embedded virtualization. Inf. Media Technol. **8**(1), 1–17 (2013)
8. Schwäricke, G., et al.: A real-time virtio-based framework for predictable inter-VM communication. In: 2021 IEEE Real-Time Systems Symposium (RTSS), Dortmund, DE, pp. 27–40 (2021). https://doi.org/10.1109/RTSS52674.2021.00015
9. Oliveira, A., Martins, J., Cabral, J., Tavares, A., Pinto, S.: TZ- VirtIO: enabling standardized inter-partition communication in a trustzone-assisted hypervisor. In: 2018 IEEE 27th International Symposium on Industrial Electronics (ISIE), Cairns, QLD, Australia, pp. 708–713 (2018). https://doi.org/10.1109/ISIE.2018.8433781
10. Dong, P., Jiang, Z., Burns, A., et al.: Build real-time communication for hybrid dual-OS system. J. Syst. Arch. **107**, 101774 (2020). https://doi.org/10.1016/j.sysarc.2020.101774
11. Dong, P., Burns, A., Jiang, Z., Liao, X.: TZDKS: a new trustzone-based dual-criticality system with balanced performance. In: 2018 IEEE 24th International Conference on Embedded and Real-Time Computing Systems and Applications (RTCSA), Hakodate, Japan, pp. 59–64 (2018)

Product Feature and Emotional Polarity Identification Method Based on ABSA Technology

Lizhuo Niu[1], Ming Cheng[2], Pan Wang[1](✉), Sichao He[1], Rui Wang[1], and Shijian Liang[1]

[1] China Automotive Data (Tianjin) Co., Ltd., Tianjin 300000, China
{niulizhuo,wangpan2015,hesichao,wangrui, liangshijian}@catarc.ac.cn
[2] China Automotive Data Co., Ltd., Tianjin 300300, China
chengming@catarc.ac.cn

Abstract. There is a massive amount of review data on e-commerce platforms, from which product designers can identify product feature keywords and corresponding emotional polarity, which can quickly understand users' satisfaction with product features. This research uses aspect-based sentiment analysis (ABSA) technology to construct the BERT + Bilstm + CRF model and BERT fine-tuning model, and uses a small amount of review data for model training. Based on the F1 value of the model, a more effective model is selected to automatically extract the product feature keywords and corresponding emotional polarity information evaluated by users from the massive review data, in order to assist designers in quickly understanding users' satisfaction with each product feature and determining the direction of product iterative design, thereby improving user satisfaction.

Keywords: ABSA technology · BERT + Bilstm + CRF model · BERT fine-tuning model · Online comment data · User requirement analysis

1 Introduction

In the current internet environment, users can post their feelings after using products on various social and e-commerce platforms. Product designers can use the review data to understand users' satisfaction of product features better, optimize design, and improve user satisfaction. However, the large amount of review data makes it difficult for designers to manually process it within a limited period of time. Therefore, in order to automate the processing and fast parsing of review data, domestic and foreign researchers have conducted extensive research on information quality analysis, feature recognition, and sentiment analysis in recent years.

There are issues with online review data such as user browsing reviews and incorrect review objects. Otterbacher, J [1] obtained 22 quantitative indicators for data quality analysis based on review data on Amazon, which serve as the basis for subsequent

G. Jeon and X. Kong (Eds.): ACIIBD 2024, LNICST 616, pp. 118–126, 2025.
https://doi.org/10.1007/978-3-031-91853-7_13

designers to use such data for analysis. With the development of machine learning technology, the evaluation of data quality is regarded as a classification problem. Chen, CC et al. [2] divided reviews into five categories to test the quality of product reviews. Hong, Y et al. [3] constructed a binary usefulness based review classification system using user preferences as features to automatically partition review data.

Some scholars have focused their research on identifying product features in data. Yang C C et al. [4] directly identified and classified product features using Bayesian technology. Jakob, N [5] selected multiple features such as characters, part of speech (POS) labels, and the shortest dependent path, and input them into the conditional random field to detect feature targets. Ma, T et al. [6] considered that the majority of review data is irregular and informal, and adopted the center theory to uniformly extract explicit and implicit opinion goals from reviews of news.

The task highly related to feature recognition is sentiment analysis, which involves both text level and phrase level. Hu, X et al. [7] proposed a supervised learning method for handling noisy and short text (SANT). Cataldi M et al. [8] utilized word dependency relationships and analyzed feature level emotional polarity based on dependency tree algorithms. Tu Haili et al. [9] used a language platform developed by the School of Computer Science at Harbin Institute of Technology to perform dependency syntax analysis and extract emotional keywords and polarity for each attribute. Liu T et al. [10] introduced deep belief networks (DBNs) to discover more abstract emotional features in review texts. Ding Y et al. [11] used rule-based unsupervised methods to create auxiliary labels and neural network models to learn hidden representations suitable for different domains.

Based on the above research status, this study proposes a method of product feature and emotional polarity identification based on aspect-based sentiment analysis (ABSA) technology. Firstly, filter the crawled review data automatically to obtain high-quality data. Then, use a small amount of high-quality data to train BERT + Bilstm + CRF and BERT fine-tuning models. Finally, select the model with higher F1 value to extract product features and corresponding emotional polarity from high-quality review data automatically, helping product designers obtain user satisfaction with product features from massive review data quickly, and assist in optimizing design.

2 Research Framework

The input for this study is high-quality review data. The output is automatically extracted product feature keywords and corresponding emotional polarity information from high-quality review data, which can help product designers understand usage experience of users quickly, assist in optimizing design, and improve user satisfaction. The framework of research is shown in Fig. 1.

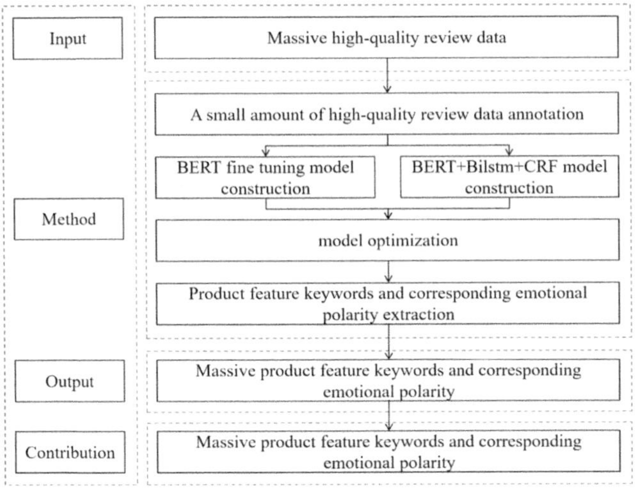

Fig. 1. Framework of research

3 Methodology

3.1 High Quality Review Data Annotation

Extracting specific product feature keywords and corresponding emotional polarity from review data can be abstracted as an aspect-based sentiment analysis (ABSA) task [12], which essentially belongs to sequence annotation problems and adopts BIO or BIOES annotation methods. Currently, machine learning methods and deep learning methods are widely used [13].

Randomly select a certain amount of data for manual annotation for model training. Due to the need for the model to simultaneously extract keywords and corresponding emotional polarity, emotional polarity labels need to be attached during the annotation process. Among them, the emotional polarity of product feature keywords is divided into positive and negative directions, represented by "P" and "F" respectively. An example of annotation is shown in Table 1:

Table 1. Example of BIO annotation process

Meaning of chinese annotations: Good appearance, loud noise									
Token	外	观	好	看	，	噪	音	大	。
Label	B-P	I-P	O	O	O	B-F	I-F	O	O

3.2 Model Construction

BERT Fine-Tuning Model

The Bidirectional Encoder Representations from Transformers (BERT) model [14] has achieved significant breakthroughs in natural language processing (NLP) technology. The model consists of two stages: pre training and fine tuning. The pre training step has been completed by google during the model building process, and only an additional output layer needs to be added for fine-tuning to achieve the best state of the art performance in various downstream tasks. The structure is shown in Fig. 2.

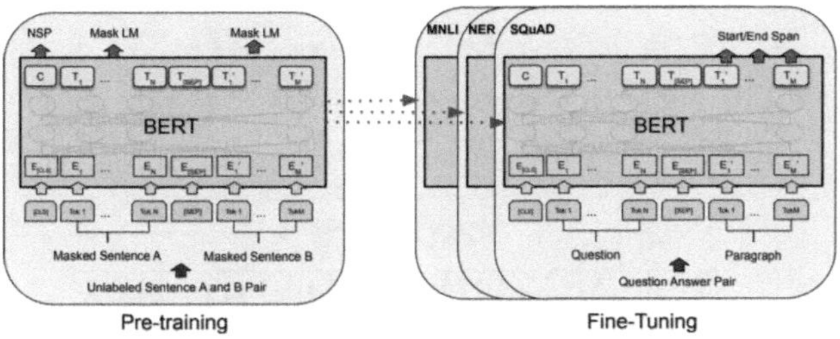

Fig. 2. BERT model structure

Considering that the BERT model can significantly improve its performance when fine-tuning downstream tasks, even if the dataset is very small, it is widely used in sequence annotation problems [15]. The BERT model can be divided into $\text{BERT}_{\text{base}}$ and $\text{BERT}_{\text{large}}$. Due to the use of Chinese review data in this study, the Chinese pre training model $\text{BERT}_{\text{wwm-ext}}$ published by the joint laboratory of Harbin Institute of Technology and iFlytek was selected, which belongs to the same base model. Based on the python language, by calling the BertForTokenClassification class in the BERT pre training model under the Transformers library in the pytorch framework, a small amount of review data is used to fine tune BERT.

BERT + Bilstm + CRF Model

With the improvement of computational power and the introduction of word embedding, deep learning algorithms such as Recurrent Neural Network (RNN) are commonly used for sequence annotation tasks. However, this method have two types of drawbacks. Firstly, they cannot predict the current character token based on the previously predicted labels, which may lead to the possibility of invalid predicted label sequences; Secondly, neural networks such as RNN are limited to connecting previous information over short distances, and there is a possibility of missing key information over long distances, leading to the possibility of incorrect prediction of label tokens.

In response to these two types of defects, this research adopts the Conditional Random Field (CRF) method, which includes transfer features that consider the order between output labels, effectively solving the first type of problem; Adopting the Bi directional Long Short Term Memory (Bilstm) method [16] can better capture contextual information and bidirectional semantic dependencies in sequences, effectively solve the second type of problem, and improve the accuracy of annotation. In the Bilstm model, the unique encoding of character tokens is generally used as input. The BERT model, which is pre trained on large-scale datasets, has a word embedding layer that provides contextual semantic information, which is very beneficial for capturing complex semantic information in sentences. Therefore, utilizing the word embedding layer provided by BERT as input to the Bilstm model has good potential for improving performance. This research constructs a BERT + Bilstm + CRF model, with the specific structure shown in Fig. 3:

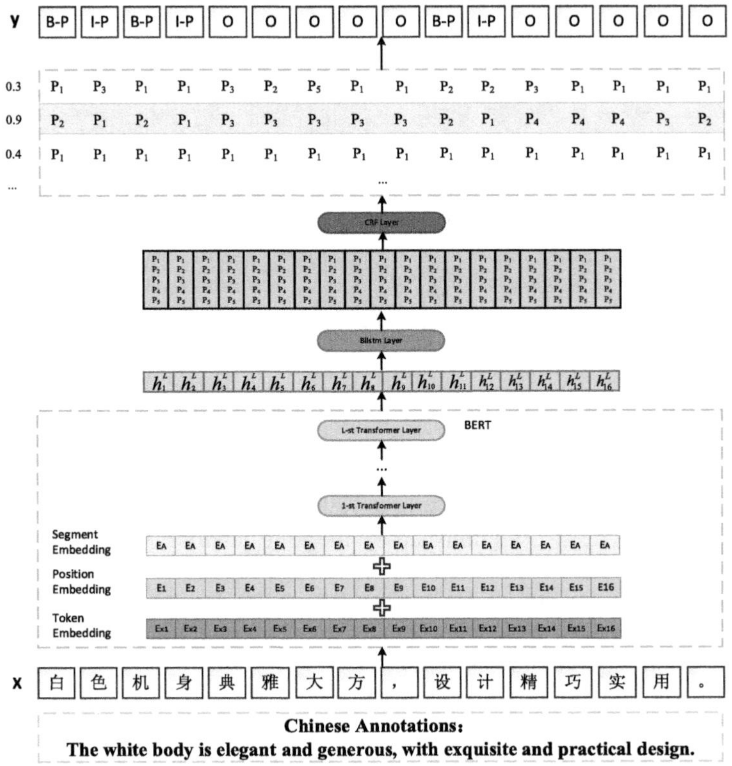

Fig. 3. BERT + Bilstm + CRF model architecture design diagram

3.3 Model Training and Optimization

Randomly divide the annotated data into 5 mutually exclusive subsets, and divide them into training and testing sets in a 4:1 ratio, respectively, for model training and model performance testing. Using the 5-fold cross validation method, take the average predicted value as the final result of model performance. Finally, based on the model performance indicators F1 value, the optimal model is selected as the model evaluation index.

3.4 Automatic Extraction of Requirement Keywords and Emotional Polarity

Using the selected optimal model, identify and extract keywords of product feature and corresponding emotional polarity from high-quality review data.

4 Case Study

This study crawled 25676 review data from a certain brand of sweeping robot on an e-commerce platform to verify the effectiveness of the proposed method. Firstly, from the 22203 high-quality review data filtered out using text classification technology, 2000 were randomly selected for annotation. Secondly, the BERT fine-tuning model and the BERT + Bilstm + CRF model were divided into training and testing sets in a 4:1 ratio using a 5-fold cross validation method.

In the BERT pre training model, set the training epoch to 3, the training batchsize to 5, and saved a loss every 30 steps. The model was saved every 200 steps and reached the maximum number of steps in each Epoch round. Therefore, a total of 7 models were saved, namely $t_{total} = 200, 288, 400, 576, 600, 800, 876$. Finally, the average value of the model indicators in the 5 fold cross validation was taken as the basis for evaluating the performance of the model, as shown in Fig. 4.

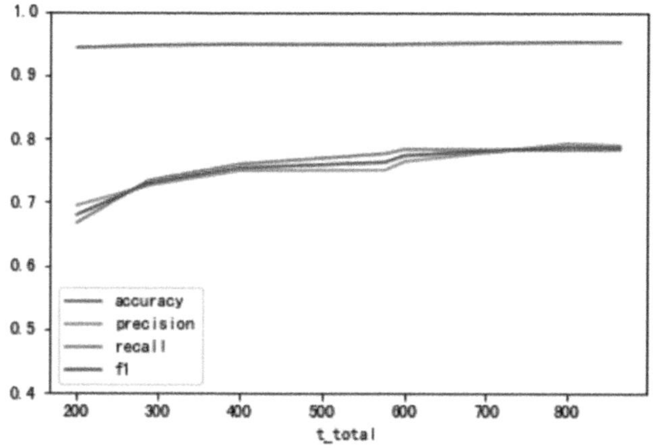

Fig. 4. Value of each indicator about the BERT fine-tuning model

From Table 2, it can be observed that the F1 value of the model was optimal at $t_{total} = 800$. Therefore, the model with $t_{total} = 800$ was selected as the final model after pre training and fine-tuning for BERT.

In the BERT + Bilstm + CRF model, a layer of CRF was added to the neural network layer constructed using the keras deep learning framework. Among them, the training epoch was set to 100, and the training batchsize was set to 32. For the parameters in the optimizer, set the learning rate to 0.001, beta_1 to 0.9, beta_2 to 0.99. Finally, the average value of the 5-fold cross index was taken as the basis for evaluating the performance of the final model, as shown in Fig. 5.

Table 2. BERT fine-tuning model indicators

T-total	Accuracy	P	R	F1
200	0.943264	0.694585	0.667397	0.680720
288	0.947001	0.726037	0.733813	0.729904
400	0.948893	0.749867	0.760349	0.755072
576	0.948584	0.751073	0.777026	0.763829
600	0.949598	0.764734	0.783729	0.774115
800	0.952787	0.793097	0.782898	0.787964
876	0.952697	0.789982	0.783479	0.786717

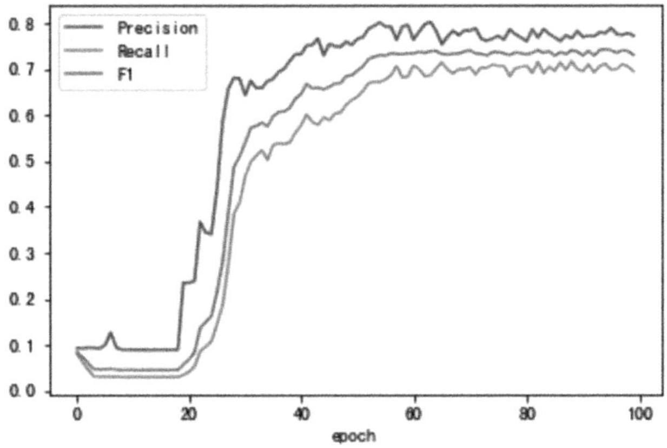

Fig. 5. Value of each indicator about the BERT + Bilstm + CRF model

When epoch = 70, the model training tended to be stable. When epoch = 96, the F1 value of model reached its maximum. Therefore, a model with epoch = 96 was selected as the final model. The details are shown in Table 3.

Table 3. Test set metrics

Epoch	Accuracy	P	R	F1
96	0.982439	0.777895	0.711941	0.743458

By compared the optimal indicators of the BERT fine-tuning model and the BERT + Bilstm + CRF model, the BERT fine-tuning model was ultimately selected to extract product feature keywords and emotional polarity from the review data. The comparison of specific indicators of the model is shown in Table 4.

Table 4. Comparison of Model Indicators

Model	Accuracy	P	R	F1
BERT fine-tuning	0.95	0.79	0.78	0.79
BERT + Bilstm + CRF	0.98	0.78	0.71	0.74

Used the trained BERT fine-tuning model to automatically extract product feature keywords and corresponding emotional polarity from 22203 data. Through statistics, it was found that 18621 pieces of data mentioned obstacle avoidance function, of which 17311 pieces expressed positive emotions towards obstacle avoidance function, indicating that the majority of users were satisfied with this function. There are 5877 pieces of data mentioning the floor dragging function, of which 5831 pieces express negative emotions towards it, indicating that most users had low satisfaction with this function. Therefore, product designers should focus on optimizing the mopping function when iteratively upgrading the sweeping robots under this model.

5 Conclusions

The purpose of this study is to automatically extract product feature keywords and corresponding emotional polarity from massive review data, assist designers in quickly understanding user satisfaction with product features, and clarify the direction of product iterative design. This study used 22203 pieces of data from a certain sweeping robot for validation. By constructing BERT fine-tuning model and BERT + CRF + Bilstm model, 2000 pieces of data were used for model training. Finally, the BERT fine-tuning model with a higher F1 value was selected to automatically extract product features and corresponding emotional polarity from 22203 data points, helping product designers understand the product features and user experience that each user is concerned about, and to identify product features that need to be optimized in iterative design.

This research method is also applicable to the recognition of feature keywords and corresponding emotional polarity for other single or multiple products. Therefore, further research will be conducted on the review data of product families in the future to assist designers in clarifying the evolutionary design path of product groups.

Acknowledgment. The research was supported by a project C03 of the Major Science and Technology Project of China Automotive Center: "Research and Equipment Development of Multilayer Safety and Reliability Testing Technology for Power Batteries" (No. ZX23240001).

References

1. Otterbacher, J.: Helpfulness' in online communities: a measure of message quality. In: Proceedings of the SIGCHI Conference on Human Factors in Computing Systems, pp. 955–964 (2009)

2. Chen, C.C., Tseng, Y.D.: Quality evaluation of product reviews using an information quality framework. Decis. Supp. Syst. **50**(4), 755–768 (2011)
3. Hon,g Y., Lu, J., Yao, J., et al.: What reviews are satisfactory: novel features for automatic helpfulness voting. In: Proceedings of the 35th International ACM SIGIR Conference on Research and Development in Information Retrieval, pp. 495–504 (2012)
4. Yang, C.C., Wong, Y.C., Wei, C.P.: Classifying web review opinions for consumer product analysis. In: Proceedings of the 11th International Conference on Electronic Commerce, pp. 57–63 (2009)
5. Jakob, N., Gurevych, I.: Extracting opinion targets in a single and cross-domain setting with conditional random fields. In: Proceedings of the 2010 Conference on Empirical Methods in Natural Language Processing, pp. 1035–1045 (2010)
6. Ma, T., Wan, X.: Opinion target extraction in Chinese news comments. In: Coling 2010: Posters, pp. 782–790 (2010)
7. Hu, X., Tang, L., Tang, J., et al.: Exploiting social relations for sentiment analysis in microblogging. In: Proceedings of the Sixth ACM International Conference on Web Search and Data Mining, pp. 537–546 (2013)
8. Cataldi, M., Ballatore, A., Tiddi, I., et al.: Good location, terrible food: detecting feature sentiment in usergenerated reviews. Soc. Netw. Anal. Min. **3**(4), 1149–1163 (2013)
9. Haili, T., Xiaobo, T., Li, X.: Research on user requirements mining model based on online reviews. J. Intell. **34**(10), 1088–1097 (2015)
10. Liu, T., Li, M., Zhou, S., et al.: Sentiment classification via l2-norm deep belief network. In: Proceedings of the 20th ACM International Conference on Information and Knowledge Management, pp. 2489–2492 (2011)
11. Ding, Y., Yu, J.F., Jiang, J.: Recurrent neural networks with auxiliary labels for cross domain opinion target extraction. In: AAAI, pp. 3436–3442 (2017)
12. Peng, H., Xu, L., Bing, L., et al.: Knowing what, how and why: a near complete solution for aspectbased sentiment analysis. In: AAAI, pp. 8600–8607 (2020)
13. Shudong, C., Xiaoye, O.: Overview of named entity recognition technology. Radio Commun. Technol. **46**(03), 251–260 (2020)
14. Devlin, J., Chang, M.W., Lee, K., et al.: Bert: pre-training of deep bidirectional transformers for language understanding. arXiv preprint arXiv:1810.04805 (2018)
15. Li, X., Bing, L., Zhang, W., et al.: Exploiting BERT for end-to-end aspect-based sentiment analysis. In: Proceedings of the 5th Workshop on Noisy User-generated Text (W-NUT 2019) (2019)
16. Graves, A., Schmidhuber, J.: Framewise phoneme classification with bidirectional LSTM and other neural network architectures. Neural Netw. **18**(5–6), 602–610 (2005)

Development of an Intelligent Inspection and Analysis System for Special Equipment Report Sampling

HuaPing Ren[1], ZhangJian Wang[1], QingKun He[2](✉), Heng Luo[1], and ChunYu Yan[1]

[1] Key Laboratory of Electromechanical Equipment Security in Western Complex Environment for State Market Regulation, Chongqing Special Equipment Inspection and Research Institute, Chongqing 401121, China

[2] School of Electrical Engineering, Chongqing University of Science and Technology, Chongqing 401331, China

2023204034@cqust.edu.cn

Abstract. The inspection management of special equipment is governed by national standards. The quality of special inspection reports directly impacts the authenticity, reliability, and national credibility of the inspection reports for special equipment. Due to the diverse types of special inspection equipment, complex inspection processes, and varying levels of inspection personnel, the standardization and reliability of inspection reports have not been well monitored. This paper develops an intelligent report sampling inspection system based on the Internet, deep learning intelligent core modules, front-end and back-end deployment, and MySQL database. The front-end is developed using VUE, and the back-end is deployed using SpringBoot. The design of the deep learning intelligent core modules includes a handwritten signature authenticity determination module based on improved GANomaly and a sampling result classification module based on HR-Net. Preliminary operational results of the system indicate that it can quickly obtain data from third-party systems, effectively determine signatures, and has a strong ability to distinguish text. The system significantly improves the inspection level of special equipment inspection reports.

KeywordS: Special inspection equipment · Inspection report · Deep learning · Intelligent module

1 Background

The Chongqing Special Equipment Testing Research Institute is the largest specialized equipment inspection and research institution in the southwestern region of China. Every year, it needs to carry out manufacturing supervision inspections, periodic inspections, and other inspection work on more than ten categories and dozens of types of special equipment, including pressure vessels, pressure pipelines, boilers, elevators, lifting

G. Jeon and X. Kong (Eds.): ACIIBD 2024, LNICST 616, pp. 127–137, 2025.
https://doi.org/10.1007/978-3-031-91853-7_14

machinery, passenger ropeways, and large-scale amusement facilities within the jurisdiction of Chongqing Municipality. As a result, there are tens of thousands of inspection reports issued annually. These inspection reports are completed by their respective inspection departments. A specific inspection report consists of three levels: inspection items completed by inspectors, inspection conclusions, and inspection report review and issuance by inspectors. Currently, the sampling inspection of these reports is done manually, and the main difficulties and problems encountered are:

1.1 The Workload of Manual Sampling Inspection is Extremely Large

Currently, the report sampling inspection at the Chongqing Special Inspection Institute is carried out by sampling inspectors from the inspection department and professionals from the quality department. According to the document regulations, the volume of sampled data must not be less than 5% of the total number of inspection reports for the quarter. However, currently, the number of all sampling personnel is less than 100. Completing the designated reports within the specified time range, filling in all the conclusions of the sampling, forming sampling evaluations, and addressing and processing sampling issues have become very cumbersome and difficult.

1.2 Authenticity of Handwritten Signatures

Due to the exceptionally numerous special inspection departments, some have introduced electronic reporting systems, such as those for pressure vessels, vehicle CNG cylinders, and safety valve inspections, which incorporate approved electronic signatures; however, some reports are still completed manually. Due to the large number of inspection reports and issues with the inspectors themselves, some electronic signatures have encountered problems such as being signed by others, incorrectly signed, or missed entirely. This has become a significant hidden risk in the reports. Particularly in cases of signatures made on behalf of others, it is generally difficult or impossible for reviewers and issuers to determine authenticity.

1.3 Classification of Issues in Sampling Inspection Reports

The primary purpose of report sampling is to identify problems encountered in inspection reports. Currently, there is a great deal of uncertainty in determining issues within inspection reports, such as unclear expressions of problems described by inspectors, significant discrepancies between the types of problems recorded and the standards, and improper logical consistency of the issues, among other practical problems. The presence of these risks has a profound impact on the credibility and authority of the inspection reports.

1.4 Quality Assessment of Spot-Check Reports

After a quality spot-check officer completes spot-checking for a period of time, it is difficult to make an objective and fair evaluation, and even more difficult to evaluate an inspector, a type of inspection, or an assessment of inspection risks based on historical spot-check records. Without historical data, it is impossible to conduct reasonable inference and risk control.

1.5 The Current Digital Platforms for Inspection of Special Equipment are Diverse

The digital software for special equipment is developed by multiple developers, involving several companies, research institutes, and universities. These developed platforms include C# development platform, C++ development platform, Java development platform, etc., and the databases also include Oracle, MySQL, SQLServer, and DB2, etc. Therefore, it is difficult to develop data communication and transmission, report conversion, and authentication work with the above platforms.

According to literature, there are currently no similar product research cases in China. In other fields, there are some mature cases of inspection report supervision systems at home and abroad. In the medical sector, there are some mature cases of inspection report supervision systems, such as Randox EQA [1], which assesses and compares laboratory performance by regularly sending samples and collecting laboratory results; Bio-Rad Unity Interlab [2], which provides comprehensive quality control solutions, including data management and inter-laboratory comparisons, to help laboratories monitor and improve the quality of their tests; CAP Proficiency Testing [3], a proficiency testing program provided by the College of American Pathologists, which helps laboratories verify the accuracy of their tests through testing samples and evaluating results; UK NEQAS [4], the United Kingdom's National External Quality Assessment Service, which provides quality assessment and feedback for various laboratory tests to help laboratories improve and maintain testing standards.

In the supervision system of special inspection equipment inspection reports, there are currently no relevant cases, let alone research conducted using artificial intelligence methods based on deep learning. Consequently, this paper launches research and work based on system development architecture, artificial intelligence modules, and related supporting software.

2 The Key Technologies of the System's Architecture

2.1 System Network Topology Design

According to the product functions provided by the software developers serving Chongqing Special Inspection Institute, the central platform of the current special inspection work is the business platform of the special Inspection Institute. This platform is responsible for collecting, processing, and displaying inspection work completed by various inspection subsystems, while also offering an interactive window for clients submitting items for inspection. The involved inspection subsystems include the boiler subsystem, pressure vessel subsystem, gas cylinder subsystem, safety valve subsystem, among others. Given the characteristics of cross-domain work by various systems, the network topology structure designed for this system is shown in Fig. 1.

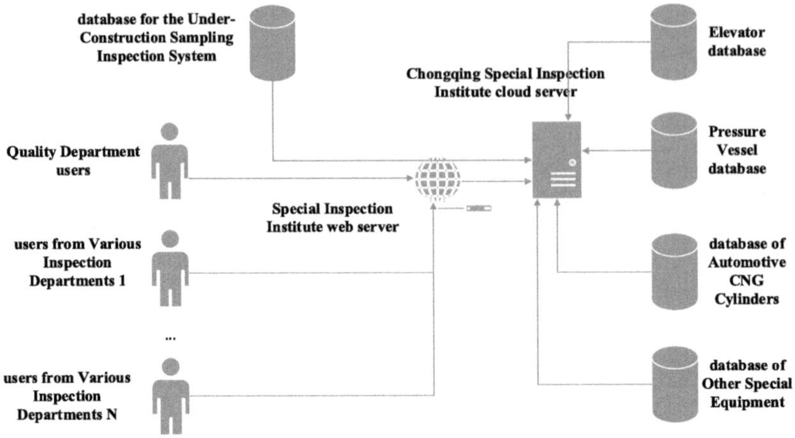

Fig. 1. System's network topology diagram

This system uses its own server, using MySQL database. Tianxin is the provider of the testing platform of the Special Inspection Institute. The gas cylinder server is provided by the gas cylinder inspection subsystem service provider, and the safety valve server is provided by the safety valve subsystem service provider. There are also some subsystems not included in the mentioned topology diagram. In the network topology design of this system, considering data encryption and the confidentiality of inspection reports, a VPN architecture has been adopted. The VPN communication is provided by a comprehensive service system from a third-party service provider, which can ensure that users at the institute level, users from various departments, and experts from external units can securely log in and operate under a unified VPN service.

2.2 Design of Data Interface Standardization

Due to the variety of special inspection equipment, as well as their categories and inspection standards, the inspection data designed by the service providers of the various inspection subsystems also vary. In response to this issue, our system conducts a comprehensive assessment of the inspection reports for all types of special inspection equipment. Moreover, it standardizes the sample format of each equipment's inspection report. This standardized design needs to be provided to all software service providers. The structural standardization design of this system is as follows:

The specifications for the interface fields are as shown in Table 1.

The interface field table has been generalized. The inspection parameters cover all the parameters that need to be spot-checked. Each subsystem fills in according to its own situation as appropriate; the personnel includes inspectors, auditors, and issuers, and the number can vary.

RESTful API: Considering that other service providers use different types of development platforms, this system is designed using a unified RESTful style. The main API designed include adding and updating inspection report records. The advantages of APIs designed in this style include:

Table 1. Interface Field Table

Field Name	Report Number	Category	Type	Inspection Conclusion	Next Inspection Period	Inspection Parameters 1–10	Personnel
Key	Yes	Yes	Yes	Yes	Yes	Depends	Yes

Platform Independence: RESTful API are based on HTTP standards, so any platform capable of sending HTTP requests can interact with them, including various web browsers, mobile applications, and even IoT devices.

Language Independence: The client can use any programming language to construct requests and parse responses, as long as the language can handle HTTP requests and data formats such as JSON/XML.

Simple and Universal: The HTTP protocol itself is widely used in various network interactions, and most programming environments provide tools for constructing and receiving HTTP requests, which lowers the technical barrier to implementing RESTful API.

In our institute, there are still some legacy systems in operation. To ensure the reliability of inspection report uploads to our system, we have adopted a point-to-point communication model.

3 The Key Algorithms of the System

3.1 Signature Authenticity Identification Based on Improved GANomaly

Due to the presence of missed, incorrect, and pending signatures in the context of this system, this paper designs an authenticity identification based on GANomaly, that is, a normal-abnormal determination. Normal is classified as genuine signatures, while anomalies are classified as forgeries, and all non-genuine signatures (including missed signatures) are considered forgeries.

The core concept of GANomaly in authenticity identification (anomaly detection) is to determine if an image is anomalous by comparing the latent space distribution of the image after encoding. For normal data, the difference between its latent space distribution and the latent space distribution after further encoding by the encoder is relatively small. On the contrary, the error in this regard is larger for anomalous data. When the difference in the latent space distribution after two encodings exceeds a set threshold, it can be identified as anomalous data [5]. The model structure of GANomaly is shown in Fig. 2, including a generator network composed of an encoder $G_Z(x)$ and a decoder $G_D(z)$; an anomaly judgment encoder $E(x)$ that performs a secondary encoding on the reconstructed image, and a discriminator D responsible for model training and optimization. The primary goal of the generative network is to reconstruct images from the input images through the generative network and learn the latent distribution of normal samples. The discriminator D serves to judge the input and output images as false or true, respectively, determining the difference between the original and reconstructed images. By continuously identifying the differences between the original and the reconstructed images until the difference between the reconstructed image and the original image is minimized, at which point the reconstructed image is almost identical to the original image. The role of the encoder *E(x)* is to compress the reconstructed images x^ from the generative network part to obtain the latent vector z^. It compares this with the latent vector z of the original images in the generative network to determine the difference in the latent space between the original and reconstructed images, which is the core of GANomaly [6].

The model's objective function integrates three loss functions, namely adversarial loss L_{adv}, context loss L_{con}, and encoder loss L_{enc}, with f representing the function of the discriminator network's intermediate layer. During the training phase, the condition of input images is evaluated by defining a differentiation score *A(x)*. If *A(x)* exceeds a predetermined threshold, the image is deemed to be an anomaly [7]. The differentiation score function are as follows:

$$A(x) = L_{enc} = \|G_E(x) - E(G(x))\|_1 \tag{1}$$

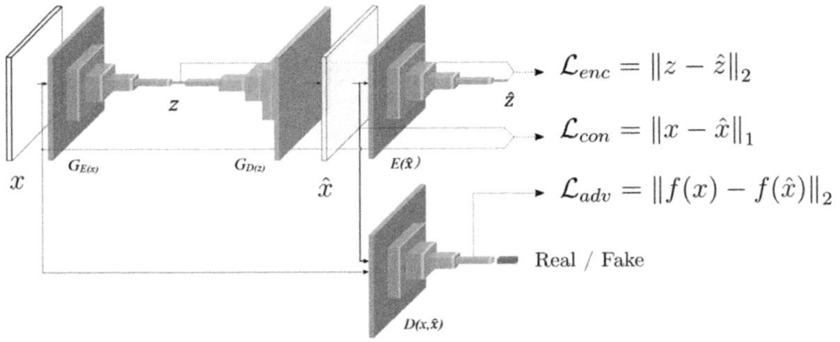

Fig. 2. GANomaly model structure

For signatures, a type of image with typically high quality and characterized as black and white bitmaps, the following optimizations have been made to the GANomaly-based authenticity algorithm:

Optimize the Image Algorithm Channels: The original algorithm needed to process data from three channels, but this paper removes two channels, retaining only one grayscale channel. That is, in the tensor of algorithm image input, the original three channels are reduced to one channel, and the corresponding convolutional operators are modified to one-dimensional operators, significantly reducing the computational load of the algorithm and increasing its efficiency.

In Multiscale Fusion, the Discriminator is Optimized: The overall framework of the multiscale feature fusion designed in this paper is shown in Fig. 3. It encompasses two different convolutional kernel fusion methods: parallel multi-branch feature fusion and serial skip connections. Both methods have the ability to extract features from different receptive fields. The principle of feature extraction in the parallel branch structure is through different sizes of convolutional kernels to gather image information within different receptive fields and then combining these different-sized kernels before passing them to the next layer of the network. Compared to parallel multi-branch feature fusion, the advantage of serial multiscale fusion lies in its ability to accommodate different levels of feature information. Adopting a multiscale fusion strategy in the classification network can help the network better extract texture features of text.

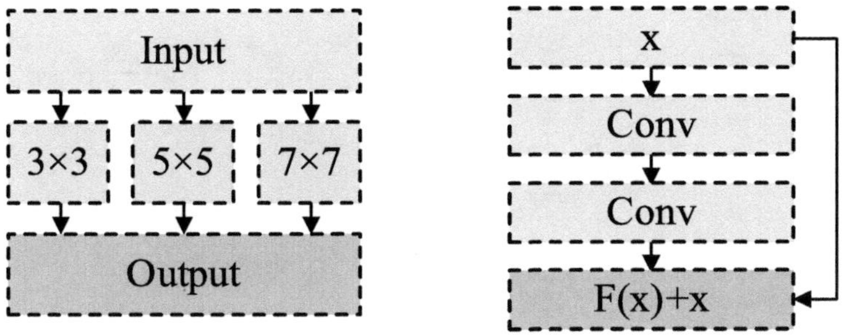

(a) Parallel multi-scale feature fusion (b) Serial multi-scale feature fusion

Fig. 3. Multi-scale feature fusion structure diagram

Inception Module: In the classification network, the structure of the convolutional layer is improved through the Inception module, which uses computational resources more effectively while enhancing the model's classification capability. The optimized convolutional architecture, without increasing the computational workload, can extract more feature information. The Inception module uses convolutional kernels of different scale sizes and includes a 1×1 convolution after each branch for alignment, thus balancing the number of channels. By establishing a parallel network structure, it enhances the network's ability to extract feature information. Hence, the parallel network uses not only the conventional 3×3 convolution but also employs multiple convolutional kernels of different scales to capture the receptive field, including 5×5 convolutions and 1×1 convolutions to extract as much feature information as closely related to the original image as possible. The innovation of Inception lies in its use of the common 3×3 convolution at the stage of extracting image features, and to optimize channel number processing, it uniquely added a parallel branch composed of 3×3 pooling and 1×1 convolution. Compared to 3×3 pooling, the 1×1 convolution and the architectural design of both 3×3 pooling and 1×1 convolution layers have the ability to help the model extract more feature information.

The structure of the Inception module is shown in Fig. 4. After the convolution operation in each branch of different sizes, the extracted abundant key image feature information is passed to the next layer in a stacked manner. To further enhance the overall performance of the model, in the branches of 3×3 convolution and 5×5 convolution, the Inception module incorporates a bottleneck layer, a 1×1 convolution, to reduce weight parameters. For the convenience of stacking branch information later, without changing the image information, the channel number of the feature information is optimized and adjusted.

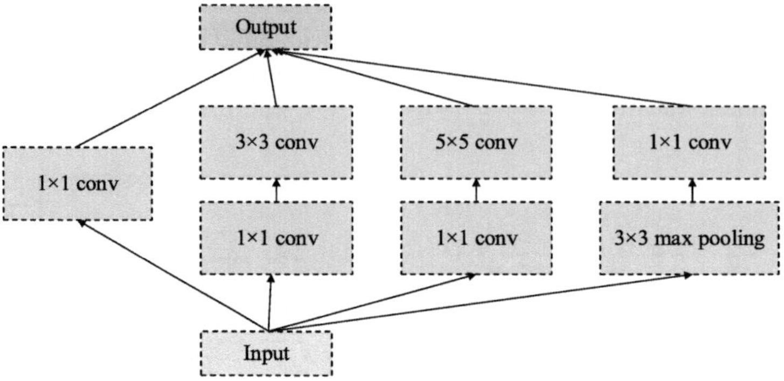

Fig. 4. Inception module structure

Adding Other Attention Modules: In the authenticity discrimination designed in this paper, other attention mechanisms are also incorporated: Single-Channel Mixed Attention (S-CBAM) module and Coordinate Attention Mechanism (CA) module.

The CBAM module consists of two submodules: the Channel Attention Module and the Spatial Attention Module. The Channel Attention Module primarily focuses on calculating attention over the channel dimension. It obtains the global statistical information for each channel by calculating global average pooling across the channel dimension, followed by a fully connected layer to yield a vector. Each element of the vector indicates the importance of the corresponding channel. Finally, the sigmoid function normalizes these importance scores to obtain attention weights between 0 and 1. These weights are then applied to the individual channels of the input feature map.

The CA module embeds position information into the channel dimensions, enabling the adaptive weighting of feature maps for each channel, thereby enhancing important features and suppressing irrelevant ones. Compared to other attention mechanisms, it does not require the computation of attention scores for all positions, which avoids substantial computational costs. This allows the network to focus on information from more regions, thereby improving the model's performance. Additionally, the CA module can be applied to images of various sizes because its attention weights are based on position information and are not fixed.

In this experiment, we randomly selected 50 inspectors from the Special Inspection Institute for our signature collection process. Each individual was asked to handwrite what they considered to be their standard signature approximately 100 times. These inspectors hailed from various departments and brought diverse cultural backgrounds, with their ages ranging from 26 to 50 years. We asked them to write their signatures at different times, using various pens and applying different levels of pressure. After gathering 5,000 signatures from these 50 inspectors, we processed the signatures through rotation, random cropping, scaling, and adjusting image brightness to both expand the dataset and ensure the generalization capability of the model.

Through the training of the model on the established dataset and after practical testing, the electronic signature authenticity detection rate designed in this paper reaches 99.98%, fulfilling the engineering requirements.

3.2 Designed an Intelligent Grading Result Comparison Algorithm Based on a Single-Channel HR-Net

In quality spot checks, one type of issue needs to be addressed seriously. That is the automatic classification of problems found in inspection reports. Currently, the Chongqing Special Inspection Institute classifies discovered issues into four categories, A, B, C, and D, based on standards, with Category A being the most serious and Category D being the least serious. In previous judgments, decisions were made manually, which presented challenges in mastering the evaluation standards. This paper introduces HR-Net [8–10] for intelligent classification processing. The main improvements made to the standard HR-Net include:

The input two-dimensional image parameters are modified to one-dimensional parameters. The one-dimensional parameters in this paper are a one-dimensional vector formulated according to the scoring standards. This one-dimensional vector adequately represents the various standards for grading. The four types of issues defined by these standards could have a one-to-one relationship or a one-to-many relationship. The vector designed in this paper is as shown in Table 2.

Table 2. Categorical Feature Vector

Vector Sub-item	Value Range
1. Completeness of Inspection Report Format	0–100points
2. Calculation Accuracy	0–100points
3. Logical Consistency in Inspection	0–100points
...	
n. Standardization of Figures	0 Non-standard, 1 Standard
...	
x. Veto Mechanism	When condition * occurs, it is directly graded as a certain level, with 0–3 corresponding to the four grades, respectively
...	

The feature vectors designed in this paper are divided into three types. The first type is a numerical type ranging from 0 to 100 points, graded by the inspectors. The second type is a chart standardizer, which is an image detector designed by other teams in this project, outputting 0 for non-standard and 1 for standard. The third type is a veto mechanism.

During the training process, historical inspection data are converted into the input vector for HR-Net, and the actual classification results are used as the target for training. After 200 epochs, the model converges. As shown in Fig. 5.

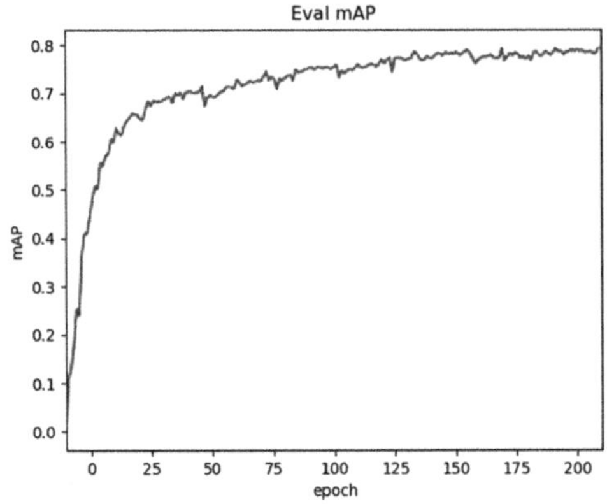

Fig. 5. The model's mAP

The final classification confusion matrix is as shown in Fig. 6.

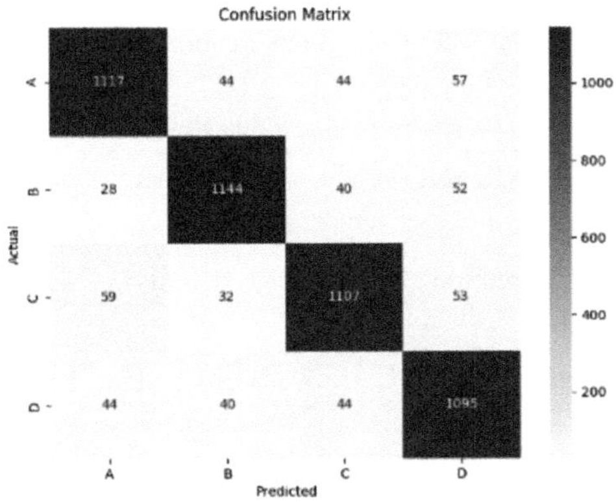

Fig. 6. Confusion matrix

4 Conclusion and Outlook

The inspection system for the Chongqing Special Inspection Institute designed in this paper, with its network architecture and software design, meets the onsite requirements. The core algorithm utilizes signature authentication based on GANomaly, as well as automatic intelligent classification of issues. Currently, it is being used satisfactorily. Future work should focus on further research into the authenticity determination in the hopes of achieving a 100% accuracy rate. The issue classification can also be further optimized.

References

1. Doherty, S., Kansanaho, G., Rodríguez, M., Adams, L., Fitzgerald, S.: Evaluation of the effect of more frequent reporting in analytical performance through the use of an international external quality assessment scheme. Clin. Chim. Acta **493**, S507–S508 (2019). https://doi.org/10.1016/j.cca.2019.03.1070
2. Tosheska Trajkovska, K., Cekovska, S., Bosilkova, G., Kostovska, I., Bogdanska, J., Topuzovska, S.: Accreditation of medical laboratories in Republic of Macedonia (2017). https://doi.org/10.1515/cclm-2017-5025
3. Rabinovitch, A.: The college of american pathologists laboratory accreditation program. Accred. Qual. Assur. **7**(11), 473–476 (2002). https://doi.org/10.1007/s00769-002-0537-0
4. Davies, K.W.: Quality assurance in clinical chemistry laboratories in the UK. Accred. Qual. Assur. **4**(1–2), 18–26 (1999). https://doi.org/10.1007/s007690050304
5. Akcay, S., Atapour-Abarghouei, A., Breckon, T. P.: GANomaly: semi-supervised anomaly detection via adversarial training. In: Jawahar, C.V., Li, H., Mori, G., Schindler, K. (eds.) Computer Vision – ACCV 2018. Lecture Notes in Computer Science, vol. 11363, pp. 622–637. Springer, Cham (2019). https://doi.org/10.1007/978-3-030-20893-6_39
6. Luo, Z., Zuo, R., Xiong, Y., Wang, X.: Detection of geochemical anomalies related to mineralization using the GANomaly network. Appl. Geochem. **131**, 105043 (2021). https://doi.org/10.1016/j.apgeochem.2021.105043
7. Liu, G., Niu, Y., Zhao, W., Duan, Y., Shu, J.: Data anomaly detection for structural health monitoring using a combination network of GANomaly and CNN. Smart Struct. Syst. **29**(1), 53–62 (2022). https://doi.org/10.12989/ss.2022.29.1.053
8. Xu, Z., Zhou, Y., Wang, S.: A novel intelligent classification method for urban green space based on high-resolution remote sensing images. Remote Sens. **12**(22), 3845 (2020). https://doi.org/10.3390/rs12223845
9. Zheng, Z., Yu, S., Jiang, S.: A domain adaptation method for land use classification based on improved HR-Net. IEEE Trans. Geosci. Remote Sens. **61**, 1–11 (2023). https://doi.org/10.1109/TGRS.2023.3235050
10. Ren, Q., Lu, Z., Wu, H., Zhang, J., Dong, Z.: HR-Net: a landmark based high realistic face reenactment network. IEEE Trans. Circuits Syst. Video Technol. **33**(11), 6347–6359 (2023). https://doi.org/10.1109/TCSVT.2023.3268062

Cultural Digital Assets in the Metaverse

Shuang Wang(✉) and Yue Cheng

China National Institute of Standardization, No.4 Zhichun Road, Haidian District, Beijing 100089, China
wangshuang@cnis.ac.cn

Abstract. The emergence of the metaverse has brought about new digital economic paradigms and given rise to a multitude of new business models. In the metaverse, cultural digital assets are the basic application scene and an important carrier of the economic ecosystem. Encouraging the development of cultural digital assets can not only promote the integration of the digital economy and the real economy, but also has great significance for the innovative development and creative transformation of the traditional Chinese culture. However, in the process of development of cultural digital assets, there are such problems like lacking of consensus and harmonized norms. As a result, this paper clarifies the concept and characteristics of cultural digital assets, establishes the classification system of cultural digital assets, and analyzes the value sources as well as development directions of cultural digital assets, so as to provide supporting for the China national strategy of digital economy.

Keywords: metaverse · digital economy · cultural digitization · digital assets

1 Introduction

Metaverse has been hailed as a new scientific and technological revolution, which is the product of a certain stage of digital economy development. If the digital economy is the penetration and integration of the Internet and the traditional economy, the metaverse is the extension and development of the digital economy in the virtual world. In 2018, the "Outline of the Development Strategy of the Digital Economy", which became the first overall strategy for the digital economy at the national level. In 2021, the "14th Five-Year Plan for the Development of the Digital Economy", which clearly put forward in the "14th Five-Year Plan" period, the digital economy should realize the overall improvement of quality and efficiency, and realize a new breakthrough in the integration of the digital economy and the real economy. As a kind of digital technology and application of the popular track, metaverse quickly to show a vigorous development in various industries. Specifically, there is a natural relationship between metaverse and culture industry. Culture metaverse is not only the origin of metaverse development, but also the field with the most metaverse application scenarios. Public cultural institutions and enterprises in many fields, such as art design, news dissemination, film and television, animation and

G. Jeon and X. Kong (Eds.): ACIIBD 2024, LNICST 616, pp. 138–147, 2025.
https://doi.org/10.1007/978-3-031-91853-7_15

game, digital reading, art performance, etc., have begun to enter into the metaverse. So various cases of cultural and blogging metaverse, performing arts metaverse, film and television metaverse, animation metaverse, gaming metaverse, non-legacy metaverse, and red culture metaverse have blossomed. Blockchain technology, as the underlying support of the metaverse, is also changing the production and circulation mode of existing cultural products in terms of both productivity and production relations, providing systematic technical support for the innovative digital culture industry ecosystem [1]. In the cultural metaverse, the traditional content production paradigm of PGC (Professionally Generated Content) has been replaced by the UGC (User Generated Content) and AIGC (Artificial Intelligence Generated Content) gradually. On the cultural content dissemination side, the metaverse breaks the barriers of space and time, realizing the seamless integration between the real and the virtual, and greatly enhancing the breadth and speed of content dissemination; on the cultural content consumption side, users can get more cultural products and services with a sense of immersion, experience and personalization in the cultural metaverse, constantly stimulating the upgrading and iteration of cultural consumption demand. With its strong vitality, the metaverse is constantly innovating and reshaping the cultural industry chain, accelerating the digital transformation of the cultural industry, and comprehensively empowering the national strategy of cultural digitization.

2 Concept of Cultural Digital Assets

2.1 Background

The rapid development of the metaverse has promoted the deep integration of the digital economy and the cultural industry, giving rise to the digitization and assetization of cultural content. The explosion of the digital artwork trading market represented by CryptoPunks has created a new industry of NFT (Non-Fungible Token) digital assets.In 2021, one NFT called "*Everydays: The First 5,000 Days*" was auctioned for $69.3 million in United States, which inspired the huge commercial potential of digital assets' huge commercial potential. According to statistics, the trading volume in February 2023 across the leading six NFT marketplaces totalled USD 1.89 billion [2]. In China, along with the heating up of the metaverse, the application of blockchain technology in the digital culture industry has greatly expanded the digital cultural products market [1].The digital collections supported by blockchain technology have been focused by many public cultural service organizations as well as cultural enterprises. According to the "Monthly Report on National Museum Digital Collections and Wenbo NFT (Cultural Creation) Distribution" released by Museum Headline, as of May 2022, as many as 52 museums nationwide have released digital collections [3]. A large number of digital cultural products trading platforms have also sprung up, and as of June 2022, there were nearly 700 digital cultural products trading platforms on line in China, according to the data of GuanRan report. The products on line on the digital cultural products trading platform are mostly based on art collections and fine art works, and also include a variety of product forms such as digital book stamps, digital music commemorative stamps, digital stamps, digital postcards, digital tickets, etc., which have greatly enriched the dissemination and consumption scenarios of the cultural industry.

Under the dual support of the national strategy of cultural digitization and digital economy, the leap from digitization to assetization of cultural content production is inevitable. Content assetization fully embodies the basic features of the digital culture industry, in which data is the key production factor, innovates the production organization and market circulation of cultural products, and provides a new feasible model for the innovative development of traditional culture. The application scenarios of cultural digital assets are diversity, such as preservation of cultural heritage, cultural tourism, trading of digital products. The "Guomai Culutre" utilized 5G and Metaverse to create a digital asset management platform for the Grand Canal, and through 3D digital interactive map, XR cultural reproduction of the Grand Canal, and Grand Canal cultural knowledge atlas, it created the immovable Grand Canal as an open-ended, sustainable digital asset that breaks down time and space constraints, and gave a second life and value to cultural assets. In addition, Nanjing, as the ancient capital of the Six Dynasties, also released the first overall solution for a digital cultural city in China based on the technologies of AI, blockchain, cloud computing and big-data, creating a digital data bank, a digital data vault and a digital data supermarket through a big data platform solution for the city's historical and cultural resources. In early 2023, at the 10th China Emerging Media Industry Convergence Development Conference, the first domestic cultural tourism immersive digital cloud platform "Cloud Tour China" was released, realizing immersive scenic tours, virtual performances, cloud cultural activities and other digital cultural tourism experiences based on the digital twin technology, and tapping the digital asset value of scenic spots, cultural relics, and IP with the new digital technology to create a new field of cultural tourism universe consumption. Specifically, on New Year's Day 2023, launch of the first national compliant digital asset secondary trading platform "China Digital Asset Trading Platform" was announced. The platform is guided by the Supervision Center of the State Administration of Radio and Television, and the official launch of the platform means that the legal nature of digital assets, transaction methods, supervision of the main body, supervision methods will be gradually clarified and the digital assets will usher in the stage of compliance, healthy and sustainable development.

2.2 The Conceptual Model

Though the growing is fast, digital assets are still a relatively new research field, there is no unified concept of digital assets, and its research also presents different views and opinions. "Assets owned by enterprises, organizations or individuals, in the form of binary electronic data, with commercial or exchange value" is considered to be the first proposed concept about digital assets [4]. But along with the development of metaverse and Web 3.0, the connotation of digital assets is constantly being expanded. Zheng and Sun believe that digital assets refer to "virtual digital tokens" under blockchain technology [5, 6]; Zhao believe that "digital assets are resources in cyberspace owned or controlled by individuals, enterprises, and the state that exist in digital form and are expected to bring economic benefits, including data assets, digital rights assets and digital currency [7]"; Gao believes that "digital assets in the narrow sense only refer to digital currency, while digital assets in the broad sense include all kinds of data generated by the information system (logistics, user flow, information flow, commodity flow, etc., which exist in electronic form and are directly related to the asset transaction) and industry data

[8]"; Wei believes that "digital assets include digital assets endorsed by national credit such as digital RMB, decentralized digital assets represented by Bitcoin, and digital assets endorsed by private credit such as game equipment and Q-coins [9]".

From the above definition, we can see that the scope of digital assets is broader than that of "data assets", which includes not only data assets, but also digital rights assets, digital currencies, virtual currencies with a certain monetary nature (such as Bitcoin), and non-monetary native digital assets (such as NFTs and digital collectibles), etc. In particular, for cultural enterprises, cultural IP (Intellecture Property), which does not exist in the form of data, is the core asset that reflects their competitiveness. In particular, for cultural enterprises, cultural IP (Intellecture Property), which does not exist in the form of data, is the core asset that reflects their competitiveness. Therefore, this paper believes that the adoption of "digital assets" can more accurately and comprehensively cover all kinds of data assets owned by cultural enterprises, as well as cultural IP and other data rights assets. Referring to the definition of cultural data assets [10], this paper defines cultural digital assets as "measurable cultural digital resources with asset ownership, cultural value and economic value", and establishes the conceptual model shown in Fig. 1.

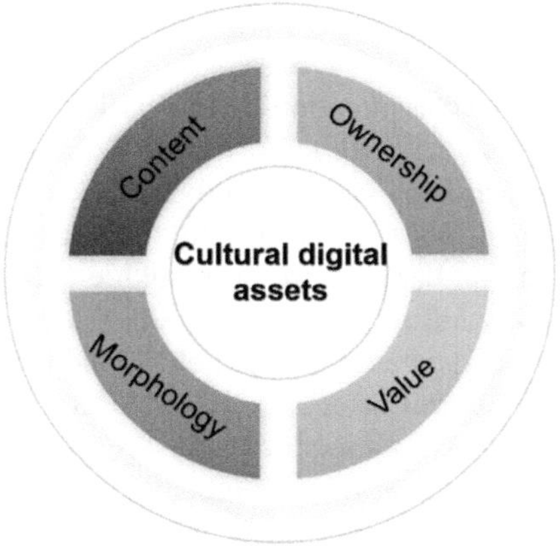

Fig. 1. conceptual model of cultural digital assets

- **Content**: The cultural digital assets should have cultural attributes, or the object which cultural digital assets lined can be attributed to the cultural industry;
- **Ownership**: The cultural digital assets should have rights attributes, i.e. they are owned by an organization or an individual;
- **Value**: The cultural digital assets should have potential or actual economic value and exchange value, which can be measured;
- **Morphology**: The cultural digital assets should be presented in digital form and do not have physical form.

As can be seen from the definition, cultural digital assets not only include assets in the form of data resources, such as data resources directly generated in the process of cultural content production, dissemination and consumption, and data resources after the digitization of physical assets, but also include digital assets presented in the form of intellectual property products. The value of cultural digital assets lies in its circulation, there is a need to integrate and exchange various types of resources, so that resources can become products for trading transactions, thus giving full play to the attributes of the economic value of cultural digital assets.

3 Classification System of Cultural Digital Assets

The classification system can clearly reveal the conceptual model and logical hierarchical relationship of cultural digital assets, and at the same time, it is also a prerequisite for cataloging management and information exchange and sharing of cultural digital assets. In this part, we first sort out the relevant classification standards about the cultural industry, and innovatively propose a classification system for cultural digital assets on this basis.

3.1 Classification Methods

Line classification and faceted classification are two basic classification methods according to the China national standard GB/T 7027-2002 *Basic principles and methods for information classifying and coding* [11]. Line classification is the classification of objects (i.e., things or concepts to be classified) according to a number of selected attributes or features are divided into a number of corresponding hierarchical classes, and arranged into a hierarchical, gradually unfolding classification system. In this classification system, the class which is divided called the upper class, and the divided categories called the lower class. Faceted classification is to consider several attributes or characteristics of a selected classified object as a number of "facets", each of which can be divided into a number of categories independent of each other. When used, the categories in these facets can be combined to form a composite category as needed.

The line classification is suitable for showing the hierarchical relationships between objects, which is like a tree. For the types of cultural digital assets are complicated, and has a hierarchical relationship between upper and lower levels, we adopted a three-layer structure based on the line classification method as illustrated in Fig. 2. It means that cultural digital assets can be categorized into a number of main categories, which can contain a number of medium categories under the main categories, and a number of sub-categories under the medium categories.

3.2 Classification System

Referring to the previous definition and relevant classification standards, the classification of cultural digital assets mainly includes digital assets in the form of data resources and digital tights assets, which adopts the line classification method and is divided into

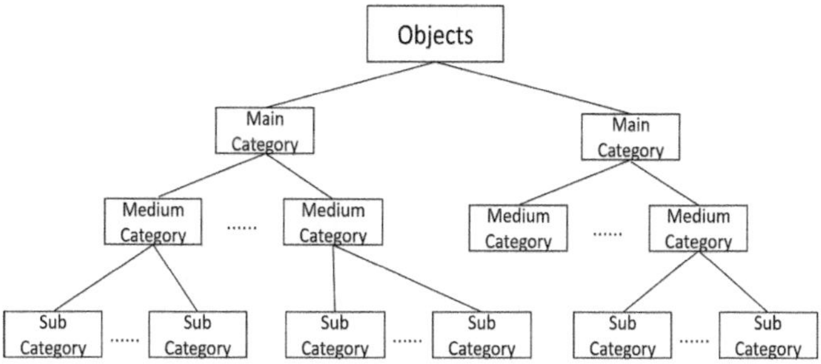

Fig. 2. Classification method of cultural digital assets

main categories, medium categories and subcategories. According to the concept of cultural digital assets, the main categories are divided into two major categories are cultural data resources and cultural rights assets, and the major categories include 6 medium categories and 37 subcategories, as shown in Table 1:

Table 1. Classification of cultural digital assets

Main Category	Medium Category	Sub Category
Cultural data resources	Digital publication	E-books
		Digital newspaper
		Digital periodicals
		Online literature
		Network education publications
		Network maps
		Digital music
		Web animation
		Web Games
		Database publications
		Mobile publications
		Other digital publications
	Audiovisual media	Broadcasting program
		Television program
		Movies
		Digital performance
		Other audiovisual media

(*continued*)

Table 1. (*continued*)

Main Category	Medium Category	Sub Category
	Visual art works	Digital crafts
		Digital artworks
		Digital photography
		Digital calligraphy and seal cutting
		Other Visual art works
	Digital design works	Electronic design drawings
		Commercial logo
		Internet advertisement
		Other Digital design works
Cultural rights assets	Digital collections	Digital tickets
		Digital music souvenirs
		Digital book collections
		Digital stamps
		Digital postcard
		Other digital collections
	IP Profile	Animation and game IP
		Literature IP
		Cultural heritage IP
		Network IP
		Other IP Profile

3.3 Application of the Classification System

The classification system is a refinement of the conceptual model, and is valuable for information catalogue, statistics and exchange. The main application of the classification includes two aspects:

(i) Asset Inclusion in Financial Statements. Currently, the digital economy is booming, *Interim Provisions on Accounting Treatment Related to Enterprise Data Resources* issued by the Ministry of Finance points out that data asset inclusion in financial statement will be implemented from January 1, 2024, which means that enterprise data resources can be effectively measured, thus transforming into valuable "assets". For the "digital assets inclusion in financial statement", the classification system is basic and necessary. Different types of the cultural digital assets, can be grouped into different financial categories, and different methods of calculating value can be applied.

(ii) Asset in Transaction Circulation. Transaction circulation is the key part to reflect the value attributes of cultural digital assets. Different types of cultural digital assets are treated differently in the process of transaction circulation. For cultural data resources, which usually exist in physical form, the transaction object is the data resource itself and the value is derived from the copyrights attached to it. For cultural rights assets, the transaction object is the physical entity to which the cultural digital assets linked, and the value is derived from the physical entity, such as the entitlement to Tickets, IP derivative products, etc.

4 Value Sources and Development of Cultural Digital Assets

4.1 Value Sources of Cultural Digital Assets

The new consumer theory holds that consumers do not derive utility from the goods they consume, but rather from the different characteristics of different goods and the services they provide. Cultural digital assets are the representative of integrating the real with the virtual world, so as to promote the real economy with the virtual world. It plays an important role in empowering the innovative dissemination of cultural IP and promoting the innovative development of Chinese outstanding traditional culture, representing the development direction of new productivity. The source of value of cultural digital assets is mainly reflected in:

(i) **Cultural IP value.** Cultural attributes are the core attributes of cultural digital assets, which, as a new type of cultural production factors, create new creative IP that are different from the existing cultural forms. Compared with traditional cultural IP, digital IP has stronger interactivity, scalability and innovativeness, and are able to better satisfy the needs of younger consumers. Through the application of digital assets, diversification of cultural IP and maximization of commercial value are realized. In the metaverse, virtual scenes and characters different from those in the real world can be created, providing new ideas for the creative development of cultural IP. Metaverse also drives new scenes of digital cultural consumption. In the metaverse, experience theaters created based on film or television IP, animation IP, literary IP, etc. have gradually become new hotspots of cultural tourism. The metaverse also provides a platform for cross-border cooperation of cultural IP, which can be deeply integrated with other fields to create more innovative and attractive cultural products.

(ii) Corresponding physical value. Cultural digital assets are the basic application scene of the virtual world and an important carrier of the economic ecosystem of the metaverse. Compared with digital collections, cultural digital assets emphasize more on the corresponding physical value sources, such as the value of physical tickets or other commemorative tickets. By combining physical commodities with digital assets and offering them on the blockchain, the unique attributes and value of the commodities can be better utilized and the offline consumption capacity of the commodities can be stimulated. Through the approach of "digital assets + cultural creation" or "digital assets + commercial IP", it can effectively promotes the combination of the digital economy and the real economy, greatly expands the consumption mode of the cultural industry, and revitalizes the traditional cultural industry with new vitality and vigor.

4.2 Development Directions of Cultural Digital Assets

After experiencing many problems brought about by the barbaric growth of digital collections, digital collections in China are gradually developing in the direction of digital asset compliance.In 2023, Beijing issued the "Implementing Opinions on Further

Accelerating the Development of the Digital Economy by Better Playing the Role of Data Elements", which explicitly proposes "exploring the construction of digital assets trading platforms based on the real underlying assets and transaction scenarios".It can be seen that China's digital collections are gradually transforming into digital assets under compliant governance, and becoming an important part of the national unified marketplace for data elements. However, there are many problems in developing cultural digital assets, such as no clear consensus on whether AIGC works are protected by copyright, or the increased difficulties in intellectual property protecting due to the ease of copying and tampering. In addition, the decentralized characteristics of the metaverse also result in the fact that the existing intellectual property system with a centralized authentication management model is not yet adapted to the mechanism of cultural digital asset production in virtual space [12]. Therefore, in the process of encouraging and developing cultural digital assets, it is also necessary to strengthen the intellectual property protection mechanism, accelerate the improvement of the construction of the relevant laws and regulations system, and encourage developing standards to form an effective regulatory mechanism. The focus should adhere to the following aspects:

Firstly, strengthening the protection of original intellectual property rights of cultural digital assets. The cultural industry is creative industry, and cultural IP is the core value source of cultural digital assets. Encouraging the creativity of digital content is the guarantee of an adequate supply of high-quality content, and to continuously enrich the content and form in the metaverse. In the process of development of cultural digital assets, it should adhere to the creativity-oriented, strengthen the encouragement for originators, and at the same time strengthen the protection mechanism for original intellectual property rights, so as to promote the development of cultural digital assets into a healthy, stable and sustainable digital asset market.

Secondly, encouraging to develop standards to guide the healthy and orderly development of the industry. Standards are the technical support for economic activities and social development, and play a fundamental and leading role in promoting the modernization of the national governance system and governance capacity. Developing standards for cultural digital assets rights confirmation, value assessment, intellectual property protection, blockchain-based transaction deposit and traceability, and the construction of transaction service platforms is of great significance for standardizing the market order of data elements represented by cultural digital assets.

Thirdly, it is need to unite other interested parties to form a good application ecology. The cultural metaverse industry has a wide range of applications and involves many subjects. It is necessary to unite industry authorities, research institutes, industry associations and enterprises and other resources to set up a cultural digital asset alliance or community, to create a community ecology of trading, sharing, communication and exchange between multiple parties. On the one hand, it can facilitate the free circulation of digital assets, so as to better realize the value of its liquidity to set up an alliance, and on the other hand, it can rely on the industry alliance to develop group standards, forming a good application ecology of joint construction by government, industry, academia, research and utilization.

5 Conclusion

The development of the digital economy is rapidly and radiating widely, and is becoming a key force in reshaping the global economic structure and changing the pattern of global competition. In the context of the national digital economy strategy, and with the important driving force of the metaverse, the cultural industry digitization has also been accelerated. In this process, a large number of cultural digital assets will inevitably be generated in metaverse, and the circulation of cultural digital assets will further accelerate the innovation and development of the cultural industry, thus promoting the development of the digital economy. Based on the conceptual characteristics and classification system of cultural digital assets proposed in this paper, it is possible for all relevant parties to form a consensus on cultural digital assets and make the circulation of cultural digital assets regulated, so as to pursue the development of cultural digital assets more healthy, standardized and sustainable.

Acknowledgements. This work was supported by the funding of "Research on Standardization of Cultural Digital Assets Management Driven by metaverse" (532023Y-10394)"of China National Institute of Standardization and the "Science and Technology Program of the State Administration for Market Regulation(No.2023MK191)".

References

1. Yu-xin, Y.U., Yu-xin, L.I.: The mechanism of blockchain technology to promote the high-quality development of digital culture industry. Shanghai J. Econ. **08**, 32–41 (2023)
2. Jayasuriya, D., Sims, A.: Not so new kid on the block: accounting and valuation aspects of non-fungible tokens (NFTs). J. Risk Financ. Manag. **16**(11), 465 (2023)
3. Wang, X., Song, Y.: A risk study of digital collections in China's museums in the context of the metaverse. View Publishing **2022**(17), 91–94
4. Dan-dan, P.A.N.: "The trinity" builds digital assets to enhance the competitiveness of the cultural industry. City Impression **12**, 166–168 (2023)
5. Zheng, P.: Case Analysis of Digital Assets Transaction Based on Blockchain Technology. Zhejiang University, p. 14 (2018)
6. Sun, Y.: Civil Law Adjustment of Digital Assets under Blockchain Technology. Hebei University of Economics and Business, p. 10 (2020)
7. Zhao, W., Liu, Q.: Digital Assets: New Infrastructure to Reconstruct the New Form of Digital Economy. People's Posts and Telecommunications Press, p. 31 (2020)
8. Ming-hua, G.A.O.: Accounting challenges of digital and platform assets in the context of digital economy - an example of group M. China Circ. Econ. **25**, 159–162 (2022)
9. Liu-Ting, W.E.I., Chu-Ming, M.O.K.: Definition of the nature and protection path of digital assets under the perspective of civil law. J. Guangxi Adm. Cadre Inst. Politics Law **37**(04), 86–94 (2022)
10. Jin, W.: The Wave of Digital Culture Economy. China Translation Publishing House, p. 223 (2022)
11. GB/T 7027–2002 Basic principles and methods for information classifying and coding, p. 2 (2002)
12. Zhang, H., Xu, Z.: Application path and risk countermeasures of metaverse empowering cultural industry innovation and development. China Internet **2023**(10), 28–32

Statistical Research of Medieval Phonology in Dunhuang Dialect

Yanhua Duan, Yonghong Li, and Xianghe Meng(✉)

Key Laboratory of Linguistic and Cultural Computing Ministry of Education, Northwest Minzu University, Lanzhou Gansu 730030, China
188142357@xbmu.edu.cn

Abstract. This paper uses the method of calculating phonological information to conduct a comparative study of the sound, rhyme, and tone of the Middle Chinese dialect with those of the Dunhuang dialect, and explores the main evolution trend of the Dunhuang dialect phonological system. The evolution of consonants is as follows: the voiced consonants are devoiced; The differentiation of lip sounds leads to the production of labiodental sounds [f]; The three groups of "Zhi, Zhuang, Zhang" differentiate and combine to produce retroflex consonants [tʂ], [tʂʰ] and [ʂ]; The three groups of differentiation and combination of "jing, jian, xiao" flow out of [tɕ], [tɕʰ] and [ɕ]; The number of zero consonants has increased significantly. The evolution of vowels is as follows: the stop consonant finals [p], [t] and [k] have disappeared; The nasal ending [m] becomes [n]; The nasal ending [n] becomes [ŋ]. The evolution of tones is as follows: the plain tone characters with clear consonants are now read as Yinping, and the plain tone characters with voiced consonants are now read as Yangping; The words with the initial tone of the consonants of the whole clear, second clear, and second muddy tones still read the initial tone; The words with a full voiced initial consonant are merged into the present tone and read as the tone; The words with initial consonants and secondary voiced consonants in the tone of the present day mainly read as Yinping; The pronunciation of the words with a fully voiced initial consonant in the entry tone is mainly Yangping.

Keywords: Dunhuang Dialect · Statistical Analysis · Dialect Model · Phonology

1 Introduction

The Qieyun represents the medieval phonology, which was lost. After the Qieyun, Guangyun became the most complete representative of the existing Qieyun system, and is the main object of research in the Middle Chinese phonology. Guangyun is useful for distinguishing sounds, correcting words, clarifying meanings, and testing words [1]. The evolution from medieval Chinese to modern dialects is very complex. Investigating the relationship between ancient and modern phonology is of great significance for studying the development patterns of Chinese dialects, conducting dialect surveys, and reconstructing medieval Chinese pronunciation. Based on the monosyllabic pronunciation table, information on the corresponding Middle Ancient Chinese pronunciation can

G. Jeon and X. Kong (Eds.): ACIIBD 2024, LNICST 616, pp. 148–155, 2025.
https://doi.org/10.1007/978-3-031-91853-7_16

be found in Guangyun, including the upper and lower characters of the anti-cut, consonants, vowels, tones, and other features such as the tone and the voice. The medieval phonological system includes 40 initial consonants, 142 finals, and 4 tones.

According to the research conducted by Liu Ling [2], Mo Chao [3], Yin Wen [4], and Wang Hongjie [5] et al., the Dunhuang dialect consists of 26 consonants (including the zero consonant), 32 vowels, and 4 tone categories. Modern Chinese has evolved from ancient Chinese. After determining the phonological system of the Dunhuang dialect, it is possible to compare the tones, rhymes, and tones of the Dunhuang dialect with those of the Middle Chinese, and analyze the evolution rules between the two. This article uses the method of calculating phonemic information to study the evolution of the sound, rhyme, and tone of the Middle Chinese dialects to the Dunhuang dialects, and explore the main evolution trend of the Dunhuang dialect phonology. The vocabulary used in this article is from the 1000-word Chinese Language Resource Survey Manual-Chinese Dialects [6]. While also referring to the "Dialect Survey Character List" [7].

2 Initial Consonant Model

The layout of Guangyun is not much different from that of Qieyun, with the four tones as the main framework, arranging the rhyming words according to the tones of Ping, Shang, Qu, and Ru.Although the initial consonants and vowels cannot be discerned from the surface of the rhyme book, by employing specific methods to categorize the words representing the same initial consonant or vowel in the upper and lower characters of the anti-tangent, one can organize the initial consonants and vowels of the medieval phonology. Ding Shengshu and Li Rong have made adjustments to the consonant system based on the division of the phonetic categories in Guangyun: separating the light labial consonants "fei, fu, feng, wei" from the labial consonants; Separate "Yu San" from the mother of the box and change its name to "Yun Mu". "Yu Si" was renamed "Yi Mu", making a total of 40 initial consonants [8].

The initial consonants of medieval Chinese pronunciation are divided into four categories: completely clear, secondarily clear, completely voiced, and secondarily voiced. These four categories were determined by Luo Changpei after consulting various names [9].The term "completely clear" refers to the unvoiced plosives, affricates, and fricatives, while the term "secondarily clear" refers to the unvoiced aspirated plosives and aspirated affricates. The term "completely voiced" refers to the voiced plosives, affricates, and fricatives, while the term "secondarily voiced" refers to the nasal, lateral, and semivowel sounds. According to 1000 words, there are 40 initial consonants in medieval Chinese, with an average frequency of 25 times. The most frequent initial consonant is [k], which appears 97 times. The number of words corresponding to each initial consonant in medieval Chinese is detailed in Table 1. The numbers in parentheses in the table represent the frequency of occurrence of the initial consonant.

The frequency of initial consonants ranges from the smallest "Chuo" to the largest "Jian", with a frequency range of 3–97, as detailed in Table 2 (The unit of frequency is "times").

The differentiation of ancient initial consonants is mainly found by examining the correspondence between the Middle Ancient Chinese phonological system of initials,

Table 1. Frequency Table of Ancient Chinese Initial Consonants

Series	Group	Clear		Voiced	
		completely clear	secondarily clear	completely voiced	secondarily voiced
Bang (162)	Bang	Bang[p]-(35)	Pang[pʰ]-(14)	Bing[b]-(27)	Ming[m]-(37)
	Fei	Fei[f]-(16)	Fu[fʰ]-(7)	Feng[v]-(15)	Hui[ɱ]-(11)
Duan (282)	Duan	Duan[t]-(29)	Tou[th]-(27)	Ding[d]-(37)	
	Ni				Ni[n]-(19) Lai[l]-(47)
	Jing	Jing[ʦ]-(31) Xin[s]-(42)	Qing[ʦʰ]-(21)	Cong[ʣ]-(18) Xie[z]-(11)	
Zhi (190)	Zhi	Zhi[ʈ]-(13)	Che[ʈʰ]-(3)	Cheng[ɖ]-(26)	
	Zhuang	Zhuang[ʧ]-(8) Sheng[ʃ]-(24)	Chu[ʧʰ]-(8)	Chong[ʤ]-(8)	
	Zhang	Zhang[ʨ]-(24) Shu[ɕ]-(18)	Chang[ʨʰ]-(11)	Chuan[ʥ]-(12) Chan[ʑ]-(17)	
	Ri				Ri[nʑ]-(18)
Jian (366)	Jian	Jian[k]-(97)	Xi[kʰ]-(41)	Qun[g]-(27)	Yi[ŋ]-(34)
	Xiao	Xiao[x]-(37)		Xia[ɣ]-(50)	
	Ying	Ying[ø]-(34)			Yun[r]-(21) Yi[j]-(25)

finals, and tones, and the Dunhuang dialect's 1000 monosyllabic word list of initials, finals, and tones. From the perspective of Middle Ancient Chinese phonology, one can also observe the different evolutionary paths of ancient initials in the contemporary Dunhuang dialect. Additionally, it is possible to explore the main trends in the differentiation of these initials. The differentiation of ancient initials in the Dunhuang dialect primarily exhibits the following five characteristics: (1) The voicing of fully voiced initials is lost, becoming unvoiced; (2) Labial consonants differentiate, resulting in the emergence of labiodental fricative f; (3) The three groups "Zhi", "Zhuang" and "Zhang" merge and differentiate into retroflex consonants tʂ, tʂʰ, and ʂ; (4) The three groups "Jing", "Jian" and "Xiao" merge and differentiate into alveolo-palatal consonants ʨ, ʨʰ, and ɕ; (5) There is a significant increase in the number of zero initials.

3 The Rhyme Model

The rhyme system of the "Guangyun" is complex, involving many concepts. In the "Guangyun," a rhyme refers to characters that have the same rhyme base (nucleus and coda) and tone, disregarding the onset, and distinguishing tones. Tang Zuofan suggests that if one does not differentiate between tones and medials, as long as the nucleus (main vowel) and coda are the same, it constitutes a rhyme category [10]. The "Guangyun" is

Table 2. Frequency Table of Middle Ancient Chinese Initial Consonants

SN	Initial Consonant	Frequency	SN	Initial Consonant	Frequency
1	Jian	97	21	Qing	21
2	Xia	50	22	Yun	21
3	Lai	47	23	Ni	19
4	Xin	42	24	Cong	18
5	Xi	41	25	Shu	18
6	Ming	37	26	Ri	18
7	Ding	37	27	Chan	17
8	Xiao	37	28	Fei	16
9	Bang	35	29	Feng	15
10	Yi	34	30	Pang	14
11	Ying	34	31	Zhi	13
12	Jing	31	32	Chuan	12
13	Duan	29	33	Wei	11
14	Bing	27	34	Xie	11
15	Tou	27	35	Chang	11
16	Qun	27	36	Zhuang	8
17	Cheng	26	37	Chu	8
18	Yi	25	38	Chong	8
19	Sheng	24	39	Fu	7
20	Zhang	24	40	Che	3

divided into four major categories based on the four tones, totaling 206 rhymes, including 57 level tones, 55 rising tones, 60 departing tones, and 34 entering tones. The level tone has a large number of characters and is divided into two volumes: upper level with 28 rhymes and lower level with 29 rhymes, which are unrelated to the modern concepts of yin level and yang level.

The rhyme categories of the "Guangyun" are distinguished by "Shu" (smooth) and "Cu" (checked) to identify the rhymes, grouping those with the same medial, main vowel, and coda into one rhyme mother based on the level, rising, and departing tones, and separating the entering tone rhymes with different codas into another rhyme mother. Thus, the 293 rhyme categories of the "Guangyun" can be classified into 142 rhyme mothers, including 91 smooth rhymes (40 yin rhymes and 51 yang rhymes) and 51 checked rhymes. The 1000 monosyllabic words in the "Chinese Language Resource Survey Handbook: Chinese Dialects" can be classified into 128 rhyme mothers, including 88 smooth rhymes (39 yin rhymes and 46 yang rhymes) and 43 checked rhymes (entering tone rhymes).

Although there are many studies on the phonetic reconstruction of the "Guangyun" sound system, this section still uses the phonetic category representation method to represent the rhymes and syllables of the "Guangyun." The rhymes of the "Guangyun" can be represented by a combination of "Rhyme", "Call" and "Rank". "Rhyme" is represented by the level and entering tone rhyme headings, with the rising and departing tones included in the level tone rhymes. For example, "Dong He Yi" represents the first rank rhyme "Dong" with a closed mouth. It can also be represented by a combination of "Division", "Call", "Rank" and "Rhyme," such as "Tong He Yi Dong," representing the first rank rhyme "Dong" of the Tong division with a closed mouth. Similarly, syllables can be represented by a combination of "Initial Consonant," "Rhyme Mother," and "Tone" using the phonetic category method, for example, "Lai Dang Kai Yi Duo Ru," where "Lai" is the initial consonant, "Dang Kai Yi Duo" is the rhyme mother, and "Ru" is the tone.

Based on 1000 characters and 128 Middle Ancient Chinese rhyme mothers, the average frequency of the rhyme mothers is 7.81 times. Looking at the types of codas, yin rhymes appear 376 times, yang rhymes 416 times, and entering tone rhymes 208 times. Among the yang rhymes, there are 9 rhyme mothers with [m] coda, appearing 56 times; 18 rhyme mothers with [n] coda, appearing 176 times; and 19 rhyme mothers with [ŋ] coda, appearing 184 times. Among the entering tone rhymes, there are 9 rhyme mothers with [p] coda, appearing 31 times; 17 rhyme mothers with [t] coda, appearing 66 times; and 17 rhyme mothers with [k] coda, appearing 111 times. The number of characters corresponding to the Middle Ancient Chinese rhyme mothers is detailed in Table 3 (The unit of frequency is "times"), which only lists rhyme mothers with a frequency of 15 times or more.

Table 3. Burden Scale of Middle Ancient Chinese Rhyme Mothers

Rhyme Mother	Frequency	Rhyme Mother	Frequency
Dang Kai San Yang	28	Zhen Kai San Zhen	18
Liu Kai San You	27	Shan Kai Si Xian	17
Yu He Yi Mo	22	Shan Kai San Xian	16
Xiao Kai Yi Hao	21	Zhi Kai San Zhi	16
Yu He San Yu	19	Jia Kai Er Ma	15
Shan He Yi Huan	19	Zhi Kai San Zhi	15
Tong He Yi Dong	19	Tong He San Zhong	15
Yu He San Yu	18	Liu Kai Yi Hou	15
Zhi Kai San Zhi	18	Geng Kai San Qing	15

By examining the correspondence between the Middle Ancient Chinese rhyme mothers and the contemporary rhyme mothers in various dialects, one can calculate the differentiation of the ancient rhyme mothers. Although there are certain patterns in the differentiation of rhyme mothers, there are also a few exceptions. This calculation will only take into account the main patterns of differentiation.The correspondence between the Middle Ancient Chinese rhyme mothers and the Dunhuang dialect rhyme mothers is primarily manifested in the following three points:(1) The plosive codas [p], [t], [k] are lost; (2) The nasal coda [m] changes to [n]; (3) The nasal coda [n] changes to [ŋ].

The Dunhuang dialect has only [n] and [ŋ] as nasal codas and lacks entering tone rhymes. The Dunhuang dialect has 32 rhyme mothers, with an average frequency of 31.25. From the perspective of rhyme mother types, 11 monophthong rhyme mothers appear 422 times, accounting for 42.2% of the total number of characters; 14 diphthong rhyme mothers appear 321 times, accounting for 32.1% of the total number of characters; 7 nasal coda rhyme mothers appear 257 times, accounting for 25.7% of the total number of characters. The rhyme mother [i] has the highest frequency, appearing 71 times, accounting for 7.1% of the total number of characters; the rhyme mother [ɚ] has the lowest frequency, appearing 3 times, accounting for 0.3% of the total number of characters. The frequency of rhyme mothers in the Dunhuang dialect is detailed in Table 4.

Table 4. Frequency Table of Rhyme Mothers in the Dunhuang Dialect

Open(12/407)	Close-mid (8/288)	Close (8/234)	Close-narrow (4/71)
ɿ(20) Si	i(71) Xi	u(70) Bu	y(27) Nv
ʅ(18) Shi			
ɚ(3) Er			
a(32) Ba	ia(10) Jia	ua(13) Kua	
ə(52) Ge	iə(32) Qie	uə(32) Tuo	yə(15) Xue
ɛ(34) Tai		uɛ(6) Guai	
ɔ(32) Mao	iɔ(20) Jiao		
æ̃(63) Nan	iæ̃(61) Qian	uæ̃(22) Guan	yæ̃(12) Quan
ei(33) Bei		uei(24) Dui	
ɤu(25) Dou	iɤu(16) Biao		
aŋ(28) Dang	iaŋ(18) Liang	uaŋ(14) Shuang	
ɤŋ(67) Geng	iŋ(60) Bing		
		uŋ(53) Zhong	yŋ(17) Yong

4 Tone Model

The "Guangyun" is divided into volumes based on the four tones, which are level (ping), rising (shang), departing (qu), and entering (ru) tone categories. By statistically analyzing the Middle Ancient Chinese tones of the characters in the "Chinese Language Resource Survey Handbook: Chinese Dialects," it is found that the level tone has the highest number of characters, reaching 375 occurrences; the departing tone has the next highest number; and the rising tone has the fewest, with 189 occurrences. The average frequency of the Middle Ancient Chinese tones is 250 times. The specific values are detailed in Fig. 1 (The unit of frequency is "times").

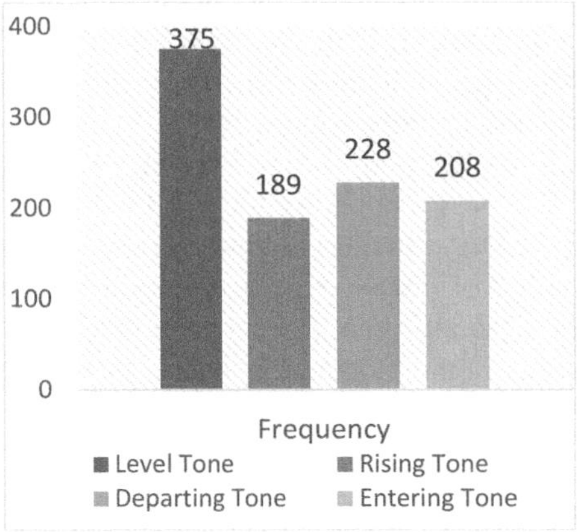

Fig. 1. Chart of Middle Ancient Chinese Tones

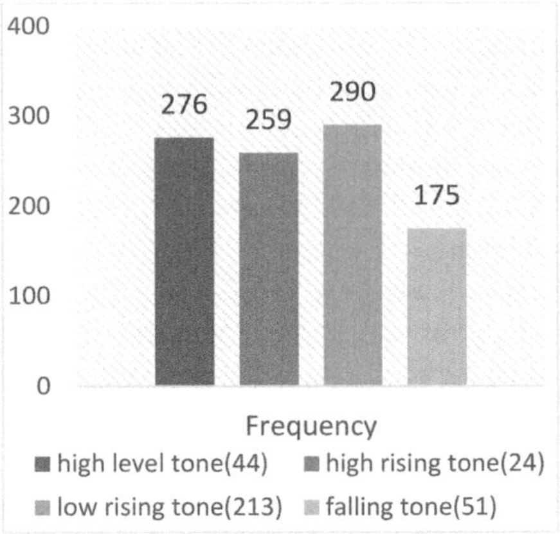

Fig. 2. Chart of Dunhuang Dialect Tones

The Dunhuang dialect has four tones, and the frequency of these tones is detailed in Fig. 2. The rising tone has the highest frequency, followed by the yin level tone, which reaches 290 occurrences; the departing tone has the lowest frequency, with 175 occurrences. In the Dunhuang dialect, examples of the yin level tone category include "po", "ke", "bu" and "ye"; examples of the yang level tone category include "pa", "ye", "yu" and "pai"; examples of the rising tone category include "kai", "hui", "ku" and "shu"; and examples of the departing tone category include "yuan", "fen", "leng" and "jing".

By examining the correspondence between the Middle Ancient Chinese tones and the contemporary tones in various dialects, one can calculate the differentiation of tones. Although there are certain patterns in the differentiation of tones, there are also a few exceptions. This calculation will only take into account the main patterns of differentiation. The correspondence between the Middle Ancient Chinese tones and the Dunhuang dialect tones is primarily manifested in the following four aspects: (1) In terms of the differentiation from the ancient level tones, the level tones are divided into yin and yang. The fully unvoiced (94.89%) and partially unvoiced (95.52%) level tones are read as yin level today, while the partially voiced (91.76%) and fully voiced (92.85%) level tones are read as yang level today. (2) In terms of the differentiation from the ancient rising tones, the fully unvoiced (96.90%), partially unvoiced (84.00%), and partially voiced (92.10%) rising tones are read as rising tones today, while the fully voiced (90.0%) rising tones are read as departing tones today. (3) In terms of the differentiation from the ancient departing tones, the fully unvoiced (95.45%), partially unvoiced (80.00%), partially voiced (95.91%), and fully voiced (92.30%) departing tones are read as departing tones today. (4) In terms of the differentiation from the ancient entering tones, the entering tones have disappeared. The fully unvoiced (88.50%), partially unvoiced (73.68%), and partially voiced (72.19%) entering tones are mainly read as yin level today; the fully voiced (77.27%) entering tones are mainly read as yang level today.

5 Conclusion

This paper adopts the method of phonological information calculation, compares the initials, rhymes, and tones of Middle Ancient Chinese with those of the Dunhuang dialect, and explores the main evolutionary trends of the Dunhuang dialect phonological system. The evolution of initials is as follows: the fully voiced initials become unvoiced; the labial initials differentiate, giving rise to the labiodental fricative f; the groups "Zhi", "Zhuang" and "Zhang" merge and differentiate into retroflex consonants [tʂ], [tʂʰ] and [ʂ]; the groups "Jing", "Jian" and "Xiao" merge and differentiate into alveolo-palatal consonants [tɕ], [tɕʰ] and [ɕ]; there is a significant increase in the number of zero initials. The evolution of rhymes is as follows: the plosive codas [p], [t], [k] disappear; the nasal coda [m] changes to [n]; the nasal codas [n] and [ŋ] remain unchanged. The evolution of tones is as follows: level tone characters with fully unvoiced and partially unvoiced initials are read as yin level today, while those with partially voiced and fully voiced initials are read as yang level; rising tone characters with fully unvoiced, partially unvoiced, and partially voiced initials still read as rising tone, while those with fully voiced initials are read as departing tone; departing tone characters with all types of initials still read as departing tone; entering tone characters with fully unvoiced, partially unvoiced, and partially voiced initials are mainly read as yin level today; entering tone characters with fully voiced initials are mainly read as yang level today.

Acknowledgment. The Natural Science Foundation of Gansu Province (Grant NO. 23JRRA1736) and the Fundamental Research Funds for the Central Universities of Northwest Minzu University (Grant NO. 31920240101) supported this research.

References

1. Xueqiong, Y.: Guangyun introduction. Bashu Book Press, Chengdu (1990)
2. Ling, L.: Dunhuang Dialectography. Lanzhou University Press, Lanzhou (1988)
3. Chao, M.: Chinese Language Resource Collection: Gansu (Volume of Phonetics). Beijing: Zhonghua Book Company (2022)
4. Wen, Y.: A study on the phonetics of the Suzhou temple dialect west of the dunhuang danghe river – with a discussion on its similarities and differences with the Dunhuang Shazhou town dialect. J. Gansu High. Normal Schools **24**(06), 43–47 (2019)
5. Hongjie, W.: An Acoustic and Perceptual Study of the Tones in Gansu Chinese Dialects. Northwest Minzu University, Lanzhou (2022)
6. Language Information Management Department of the Ministry of Education, Chinese Language Resource Protection Research Center. Chinese Language Resource Survey Handbook: Chinese Dialects. Beijing: The Commercial Press (2015)
7. Institute of Language, Chinese Academy of Social Sciences. Revised Version of the Dialect Survey Character List. Beijing: The Commercial Press (2020)
8. Xianchu, W.: An Outline of Phonology. Wuhan University Press, Wuhan (2012)
9. Luo Changpei. Introduction to Chinese Phonology [M]. Beijing: Zhonghua Book Company, 1956
10. Zuofan, T.: Phonology Courseware. Peking University Press, Beijing (1987)

A Runtime Design of WebAssembly Container for Embedded Real-Time Operating Systems

Wenzhuo Wang and Jian Li(✉)

The 32Nd Research Institute of China Electronics Technology Group Corporation, Shanghai 201808, China
ecilijian@163.com

Abstract. The environment of software deployment and DevOps tended to be lightweight and packaged for easier distribution, and container technology was introduced to meet this demand. Containers provide isolation and security policies to enhance the reliability of users' applications, but they also come with huge system overhead. For lower response latency and smaller storage occupancy, a more acceptable container solution is needed for those complex environments and edge devices. This paper proposes a lightweight container runtime based on WebAssembly. We have compared the performance with the most popular runtimes and native applications, and the results show that our runtime occupies less memory and can respond faster in certain running modes, which can meet the requirements for the application deployment on the resource-constrained devices in real-time scenarios.

Keywords: Container Runtime · WebAssembly · RTOS · Embedded Systems

1 Introduction

The centralized architecture of traditional software has gradually exposed various defects in the complex environments nowadays. Meeting the software deployment requirements for specific scenarios becomes harder. The emergence of Linux containers, such as LXC and docker [1] has brought significant convenience to application development and maintenance, as well as large-scale deployment, which formed the current container technologies discussed in many research fields.

Linux containers do the job of unpacking users' codes and dependencies from an image file, and executing the applications inside. Linux containers were executed on the application layer [2]. Multiple containers can be run in the same time and all keep isolated from the host OS. These functions rely on the Linux kernel features named "Namespace" and "Cgroups" [3]. Container runtime monitors and restricts the container resources. Because of the active state of containers in the open source community, running Linux containers on personal PCs or servers is quite easy. However, in most edge devices, especially equipment or weapons in real-time embedded scenarios [4] such as UAV or airborne systems [5], it is not feasible to use Linux and Linux containers. A number of

G. Jeon and X. Kong (Eds.): ACIIBD 2024, LNICST 616, pp. 156–164, 2025.
https://doi.org/10.1007/978-3-031-91853-7_17

edge devices still need a solution to deploy and run those small and fast applications easily, just like the containers in Linux.

This paper has designed a container runtime based on WebAssembly for embedded real-time operating systems to help with the application deployment on embedded devices. Our work includes:

- Designing an embedded container runtime that follows the OCI specification;
- Designing an execution mode of wasm module which can be managed in the container environment;

The rest of this paper will be organized as follows: Sect. 2 states the background of several key technologies and some related works. Section 3 explains the overall structure and the details of this container runtime. Section 4 evaluates some metrics and some comparisons were made with some other popular runtimes. Section 5 summarizes the paper.

2 Background and Related Works

2.1 Linux Containers and Runtime

Compared to virtual machines, using containers is a lighter way for inter-application isolation. The earliest published container product is LXC, and the current mainstream containers are Docker and Kubernetes. Harri Manninen [6] mentioned several container products under Linux embedded devices, and analyzed the performance of Docker, LXC, Balena and Flatpak.

In Toni Lammz's study [7], he analyzed some metrics such as overall latency and memory consumption when deploying applications for embedded rt-Linux systems using docker containers. The result proved that the performance of docker in embedded devices is fine, but more storage space will be needed for better utility. SungHwan Jeon [8] compared the runtime performance of four commonly used container products, including kata, firecracker, gvisor and runC. We proposed a runtime using WebAssembly technology which will be introduced later, and if differs from the general OCI container runtime such as runC.

2.2 WebAssmebly

WebAssembly (wasm for short) is a low-level code which was first proposed to solve the performance defect of JavaScript on Web [9]. Instead of writing wasm code manually, developers issues compilers (such as LLVM and Emscripten) that can use wasm as a compilation target to compile source programs written in languages such as C/C++, Rust and other languages into wasm bytecode, and put it into the wasm virtual machine (also known as engine or runtime) to interpret and run the application.

Wasm code uses linear memory model and stack-based virtual machine ensuring that the applications cannot access areas outside of the module memory. The size of the distribution module is very small, and both the wasm bytecode and machine code files are quite suitable for transportation to resource-constrained embedded devices.

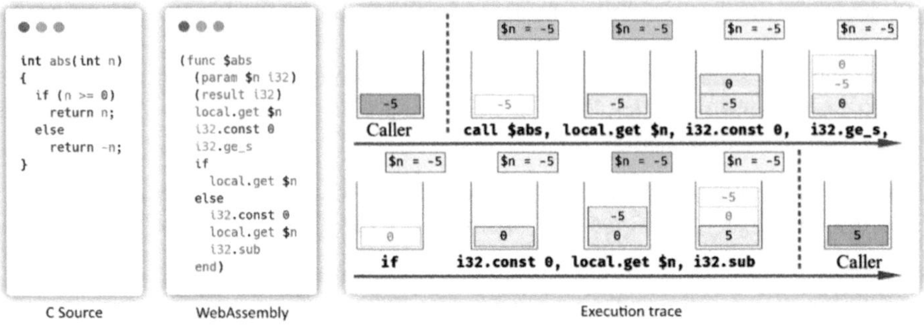

Fig. 1. WebAssembly Compilation Flow

Figure 1 gives the process from the C source code to the WebAssembly readable file and to its execution stack from left to right.

WebAssembly runs an application in an isolated environment [10]. It uses a module as the minimum distribution unit, which contains functions, global variables, tables and memory information. Zhang conducted a detailed study of the WebAssembly runtime [11], analyzing the structure and runtime characteristics of wasm, as well as the performance of more than four types of runtime, including WasmEdge, Wasmtime, and WAMR.

2.3 Embedded Real-Time Operating Systems

Unlike Linux or Windows systems, real-time operating systems (RTOS) are widely used in environments with complex time and resource constraints, and the applications usually have demanding time and performance limits [12]. The most popular RTOSs published by SourceForge in 2024 include VxWorks, FreeRTOS, RT-Thread and MbedOS [13]. Some key features have been defined in the IEEE POSIX 1003.1b,5 standard, which gives several basic functions that RTOS must provide. These are, real-time threading, inter-process communication, sync/async I/O, real-time scheduling and other mechanisms. Embedded real-time operating systems have additional requirements for the overall mirroring volume of the system and the memory consumption of the application, which also makes it difficult for some applications with more linked libraries or too large size to execute on embedded devices.

3 Runtime Design

The container runtime is generally at the bottom of the entire container architecture, acting as the cornerstone and a concrete implementation of various solutions to provide an isolated environment for users' applications. A typical container architecture is given in Fig. 2.

Container Runtime provides an interface for the container engine to create and run the containers, while from the perspective of the operating system, the container runtime is responsible for using kernel features to set attributes such as resource control groups,

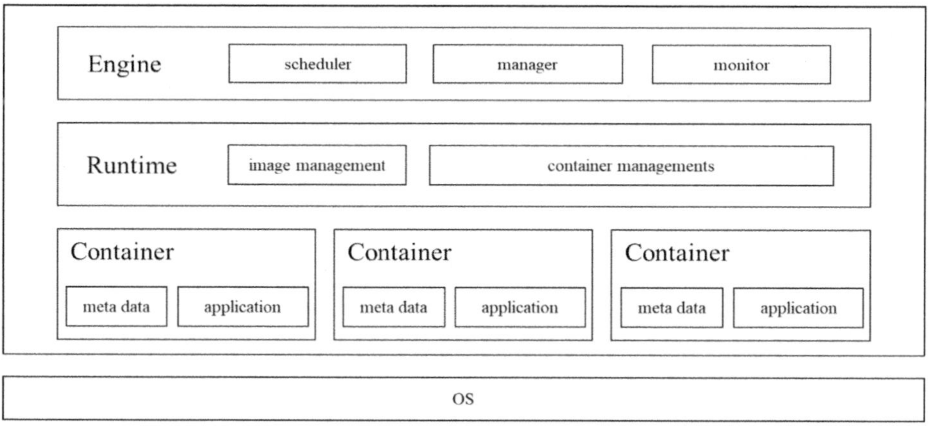

Fig. 2. Container Architecture

namespaces, and working directories for the started container process. Our container runtime does not use these features because nearly none of the RTOSs have them. Embedded RTOSs provide different system calls to manage the thread and memory, but cannot provide a lot of resource management functions like Linux. So our runtime will not use any Linux kernel-based features, but as much as possible to use the C library and Wasm API to provide various interfaces.

3.1 Overall Framework

The container runtime contains the definition of the container structure, the entire container life cycle, the management of the wasm runtime, and the execution mode wasm modules. The runtime architecture is shown in Fig. 3.

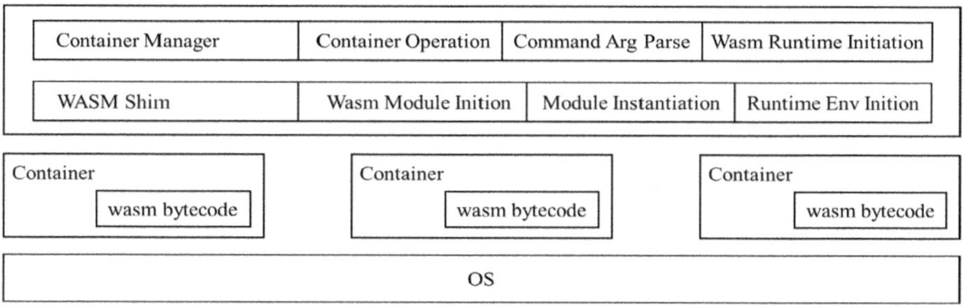

Fig. 3. Runtime Architecture

Container manager deals with the container life cycle and container operations. Figure 4 gives the state transition process of container operations and container life cycles.

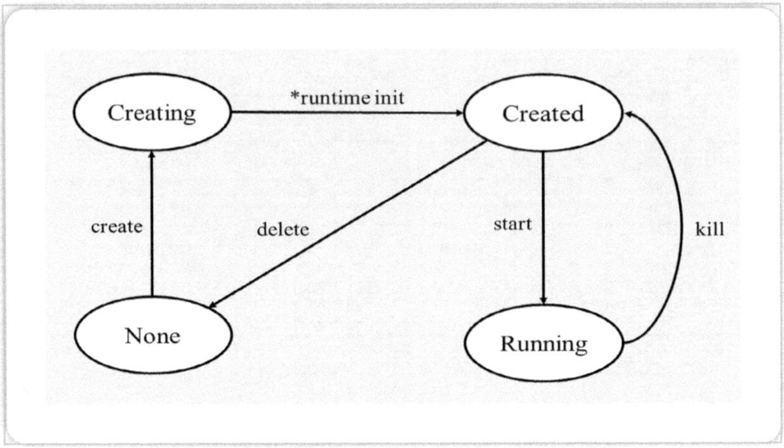

Fig. 4. Container Lifecycle

When the Container Manager receives the request, either from the command line or from the interface, it first parses the commands and parameters and generates the configuration object. Then we move to the next part to create and run a wasm container. The specific process is shown in Fig. 5:

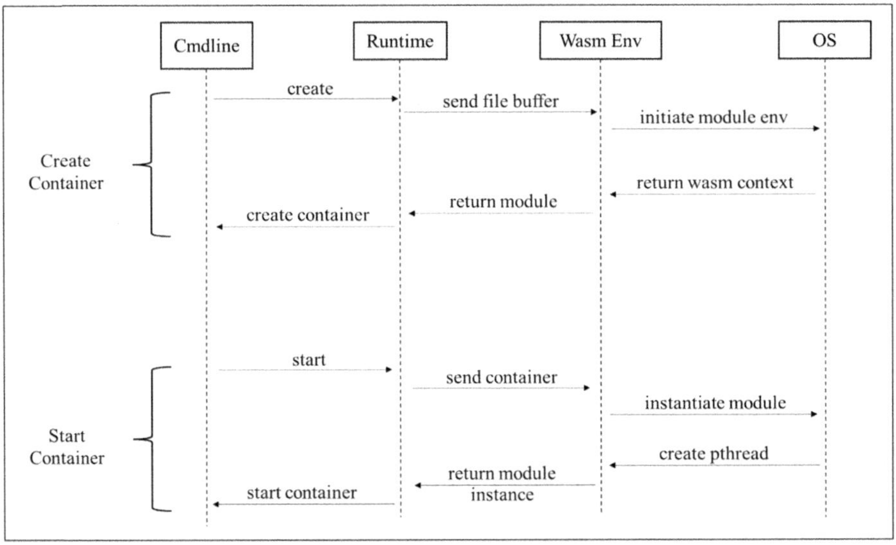

Fig. 5. Container Creation and Execution Process

In the creation phase: firstly, the running environment is initialized with the attributes parsed from the configuration. The buffer and memory space required for the execution of the wasm module will be created, and a wasm module is initialized to make a container object. The relevant information will be written into the global configuration file of the container runtime, which is located in a preset local directory.

In the execution stage: The runtime takes out the wasm configuration generated in the container config and then instantiates the wasm module. At this time, the wasm instance will be executed using the C API interface provided by wasm. The real execution is done into the code of the source program written in C or other languages.

After the above stages run successfully, the corresponding object will be returned. If any one fails, the instances will be cleaned up and an error message will be returned.

3.2 Container Management

Each container instance contains the following members.

- Configuration Parameters contains the system meta data
- Unique Identifier is used to identify each container
- Working Directory is the place to run the application
- Status saves the current status we mentioned in the life cycle
- Instance of wasm is the wasm execution context

Container Manager reads in the commands and parameters, and uses the container applications and configuration file paths to generate configuration data for the container to be created.

In the process of creating a container, the system timestamp is combined with a random number to generate a string of unique identifiers for the container, and the container working directory is generated in the global configuration directory where the read and write operations of the container will be performed. Most Linux container products choose the union file system for container image storage and working directory management, but this is something a container runtime should consider. We only need to specify a bundle directory to start running, just the same as runC or Crun.

Running any wasm module or wasm application in the container requires the initialization and instantiation of the module data structures, the allocation of memory and stack, and the setup of the runtime environment through calls to the wasm interface. This part of the work is put into container management. In fact, the process of container creation and operation is closely connected with the creation of wasm environment and the creation and operation of modules.

3.3 Container Working Mode

When a container is created, the wasm execution mode is manually specified in the command line, and the others are generated according to the default configuration.

The container runtime client provides two modes, the interpreter mode ("--interp") and machine mode ("--aot"), which create different operating environments for wasm bytecode files and aot machine code files respectively. AOT (Ahead Of Time) and JIT (Just In Time) compilation forms are often mentioned in wasm. The former compiles the source code into machine code before running, so it can have a faster execution speed. The latter one creates the wasm module which is not machine code, so a virtual machine is needed to do the work of interpretation and execution.

Applications in AOT mode execute faster, but are generally larger due to precompilation and rely on a separate compiler to compile wasm bytecode to the target platform. Applications in JIT mode execute slightly slower, but their wasm format is platform-independent, easy to distribute, and smaller in size. The choice between the two modes of operation can be set when the container is created.

4 Evaluating and Result

The test was conducted on VMware Workstation and Ubuntu 20.04 system with 4 logic cores, and the kernel is patched with Linux 5.4.0-271-rt89 for fully preempt mode to mimic the real-time environment.

The metrics to be examined include the startup time for each container individually, the execution latency of the container application and the memory usage. We choose runC, Crun, and native applications as the control groups.

4.1 Container Start-Up Time

Container startup time is a concern in most real-time scenarios, especially where a large number of containers have been deployed, such as surveillance cameras networks in urban roads or the earthquake warning devices.

The method adopted is to time the period between the initialization of the container and the start of the application execution. Since most of the container runtimes do not provide a display of this metric, it is necessary to simply modify the source code and add a timing function. We choose clock_gettime function and timespec struct instead of gettimeofday or other timing functions from Linux. The start and the end are set on the beginning of runtime main function and the line before user application is actually executed. Table 1 shows the startup time consumption during 100 startups.

Table 1. Single Container Start-up Time

Runtime	Latency(ms)		
	Min	*Max*	*Avg*
runC	278.9	1398	788.2
Crun	122.1	783.8	266.8
Our Runtime(JIT)	13.4	51.1	20.30
Our Runtime(AOT)	7.48	34.74	10.35

4.2 Execution Latency

Latency is another important metric. A simple multi-threaded test case was used here, and we execute a calculation task in a fixed cycle preset in the child thread. The execution delay was counted and returned it to the main thread. Table 2 shows the execution delay under 1000 iterations.

Table 2. Application Execution Latency

Runtime	Latency(us)		
	Min	*Max*	*Avg*
runC	55.5	10196	428.6
Crun	40.2	7726.9	451.2
Our Runtime(AOT)	55.8	7980.5	511.5
Our Runtime(JIT)	1188	9956.7	2850
Native	54.2	7427.1	499.2

4.3 Memory Usage

Also we test the memory usage when running a container. The popular products such as Docker and Kubernetes will provide interfaces such as docker stat, kubectl top and others to show the memory occupation of the container, but their metrics have a slight difference. In this paper we first use the top and free commands to view the memory usage of the process itself. But the memory of some other processes created by the runtime will be counted. We only want to focus on the runtime itself and the application thread. So pmap is the final solution we choose to trace the memory usage. Additionally, the file storage aims to find out the difference in image format. Table 3 shows the average memory usage and execution file storage usage of 20 containers at runtime.

Table 3. Memory and disk usage of a single container

Runtime	Usage	
	Memory(MB)	*Disk(KB)*
runC	42.3	899
Crun	15.1	689
Our Runtime(AOT)	8.7	104
Our Runtime(JIT)	12.4	87
Native(static-link)	9.2	1481
Native(dynamic-link)	10.4	17

The results show that the container runtime designed based on WebAssembly has a big advantage over the original Linux container runtime in terms of both startup time and space occupation, which is related to WebAssembly's good memory limiting mechanism. In terms of the execution latency of the application, the latency of AOT mode is almost close to that of the native application, while the latency of JIT mode has a bigger gap compared to the native one, but also similar to the performance of C-based Crun, exceeding the Go-based runC.

5 Conclusions

This paper presents a WebAssembly-based container runtime, which offers im-proved performance for real-time systems, and can meet the storage and memory requirements of most embedded devices. By conducting the evaluations and comparisons with native applications, RunC container applications and Crun container applications, we obtained a better experimental results. This provides a design concept for application development and deployment of equipment re-source-constrained equipment in real-time scenarios.

References

1. David, B.: Containers and cloud: from LXC to Docker to Kubernetes, Cloud Comput. **1**(3), 81–84 (2014). IEEE, https://doi.org/10.1109/MCC.2014.51
2. Da Silva, V.G., Kirikova, M., Alksnis, G.: Containers for virtualization: an overview. Appl. Comput. Syst. **23**(1), 21–27 (2018). https://doi.org/10.2478/acss-2018-0003
3. Dirk, M.: (2014) Docker: lightweight Linux containers for consistent development and deployment. Linux J. **239**, 76–91 (2014). https://doi.org/10.5555/2600239.2600241
4. Queiroz, R., Cruz, T., Mendes, J., Sousa, P., Simões, P.: Container-based virtualization for real-time industrial systems-a systematic review. ACM Comput. Surv. **56**(3), 59 (2023). https://doi.org/10.1145/3617591
5. Wen, L., Xiaofeng, L., Jianhua, M., Kaisi, F., Guo, Q.: Intelligent inspection scheme of UAV based on airborne computer. Comput. Measur. Control **30**(7), 181–186, 206 (2022)
6. Manninen, H.: Evaluation of Container Technologies for an Embedded Linux Device, Aalto University (2020)
7. Toni, L.: Feasibility of application containers In: Embedded Real-Time Linux, Tampere University of Technology (2018)
8. SungHwan, J., Simon, S., Haejin, C., Yunmook, N.: OCI runtime comparison and analysis study. In: 2022 IEEE Fifth International Conference on Artificial Intelligence and Knowledge Engineering (AIKE). Laguna Hills, CA, USA (2022). https://doi.org/10.1109/AIKE55402.2022
9. Rosenberg, A., et al.: Bringing the web up to speed with WebAssembly. Commun. ACM **61**(12), 107–115 (2018). https://doi.org/10.1145/3282510
10. Mohammed, N.H., Khaled, A.H.: WebAssembly for edge computing: potential and challenges. IEEE Commun. Stand. Mag. **6**(4), 68–73 (2022). https://doi.org/10.1109/MCOMSTD.0001.2000068
11. Yixuan, Z., Mugeng, L., Haoyu, W., Yun, M., Gang, H., Xuanzhe, L.: Research on WebAssembly Runtimes: A Survey. ArXiv, abs/2404.12621 (2024)
12. Walter, C., Phillip, A.L.: An overview of real-time operating systems. J. Assoc. Lab. Autom. **12**(1), 40–45 (2007). https://doi.org/10.1016/j.jala.2006.10.016
13. SourceForge. Best Real-Time Operating Systems (RTOS) (2024). https://sourceforge.net/software/real-time-operating-systems-rtos. Accessed 14 Jul 2024

GraphModo: An Easy-to-Use Visual Construction System of Deep Learning Model

Peng Yang, Yanyang Zhou, Liyun Chen, and Zhifeng Zeng(✉)

School of Computer Science, Zhuhai College of Science and Technology, Zhuhai 519041, Guangdong, China
zzf@zcst.edu.cn

Abstract. To address the high barriers and complexity challenges in deep learning applications, this paper proposes an innovative solution, the GraphModo system. This system effectively reduces operational difficulty through its intuitive visual interface and simplified workflow. By integrating front-end technologies based on Vue with back-end technologies in Python and Java, GraphModo enables intuitive operations such as node dragging, connections, and form-based parameter configuration, significantly lowering the technical barriers to model construction. Moreover, the system includes a rich library of classic models, provides quality and performance monitoring during the training process, and supports chart visualization, enabling users to efficiently complete modeling tasks and track training progress in real-time. GraphModo also establishes a sharing platform that encourages users to share models and configurations, further promoting exchange and collaboration. After system testing, version iterations, and performance optimization, GraphModo has achieved significant enhancements both functionality and performance, and user feedback has validated its effectiveness.

Keywords: Deep Learning · Model Building · Visual Development System · Performance Monitoring · Sharing Platform

1 Introduction

With the rapid development of artificial intelligence technology, deep learning has emerged as a focal point of research and practice due to its ability to solve complex problems. Its applications in fields such as autonomous driving, smart healthcare, smart agriculture, and security monitoring have demonstrated its immense potential. However, the widespread adoption and application of deep learning technology are not without obstacles. For beginners, the theoretical foundations of deep learning are complex, involving advanced mathematics and programming skills, which constitute a formidable barrier. Furthermore, the process of constructing, training, and optimizing deep learning models requires meticulous adjustment of numerous hyperparameters, as well as a certain level of proficiency in data processing and model tuning.

G. Jeon and X. Kong (Eds.): ACIIBD 2024, LNICST 616, pp. 165–176, 2025.
https://doi.org/10.1007/978-3-031-91853-7_18

In recent years, various frameworks have emerged to support users in graphically constructing deep learning tasks.

Weights & Biases (W&B) [1] is a machine learning platform for experiment tracking, visualization, and model management. It integrates with frameworks like TensorFlow and PyTorch but may face challenges in large-scale distributed training. Future work could improve its support for complex tasks.

Paperspace Gradient [2] is a platform for managing, tracking, and visualizing machine learning experiments, integrating with TensorFlow and PyTorch for real-time monitoring and version control. However, its performance may be limited for large-scale distributed training and complex projects.

Visdom [3] is a deep learning visualization tool for real-time monitoring and charting of training metrics. It integrates well with various frameworks but may struggle with large datasets and lacks advanced customization and multi-dimensional visualization features. Future work could address these limitations.

DragDL [4], based on the PaddlePaddle [5] framework, offers high usability through a drag-and-drop interface that abstracts the network as a dataflow graph for easy visualization. It features model performance monitoring and a library of classic models for building new ones.

Although these tools have simplified the process of constructing deep learning models to some extent, they often focus on the automation of model training while neglecting the user's intuitive understanding and control over the model-building process. This limitation prompts us to consider how to provide a more intuitive and user-friendly platform for building deep learning models.The main advantages and contributions of the system are as follows:

1. Intuitive Model Construction: Leveraging the VUE framework, GraphModo offers a user-friendly interface for constructing models through drag-and-drop nodes and form-based configurations.
2. Efficient File Export: Integrating MySQL for logic storage and a Java backend, GraphModo supports one-click export of YAML configuration files, ensuring parameter consistency and enhancing team collaboration and research efficiency.
3. Training and Evaluation Visualization: GraphModo provides a variety of visualization tools, including charts, line graphs, and scatter plots, to present training and evaluation results.
4. Community Sharing Platform: Built on Spring Boot + MyBatis, this platform allows users to share models, access classic model libraries, and collaborate, accelerating the modeling process.

The comparison results of GraphModo with existing model construction systems are shown in Table 1.

Table 1. Existing Model Construction Analysis Comparison

Features	GraphModo	W&B	DragDL	Visdom	Paperspace
Community Support	√				
YAML File Export	√				
Visual Construction	√		√		
Training Result Visualization	√	√	√	√	√

2 System Architecture

The GraphModo system is illustrated in Fig. 1.

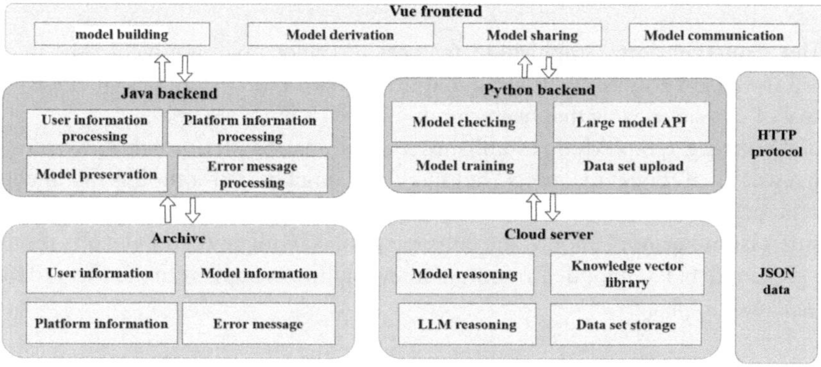

Fig. 1. Overall Architecture of the GraphModo System

The Frontend of GraphModo is developed using the Vue.js framework, which leverages declarative rendering and component-based architecture to create a user-friendly and highly interactive interface. The design of the Frontend focuses on providing an intuitive and engaging experience for users as they construct deep learning models.

GraphModo uses a hybrid architecture combining Python and Java for deep learning model construction and training. The Python Backend handles large-scale model inference and training with TensorFlow and PyTorch, offering GPU acceleration and high performance. The Java Backend uses Spring Boot for RESTful APIs and MyBatis for data persistence with MySQL. The system employs HTTP and JSON for Frontend-Backend communication.

3 System Function Design

3.1 System Overall Functionality

GraphModo is a comprehensive system that integrates deep learning model construction, model sharing, and result visualization. The overall functionality of the system is illustrated in Fig. 2.

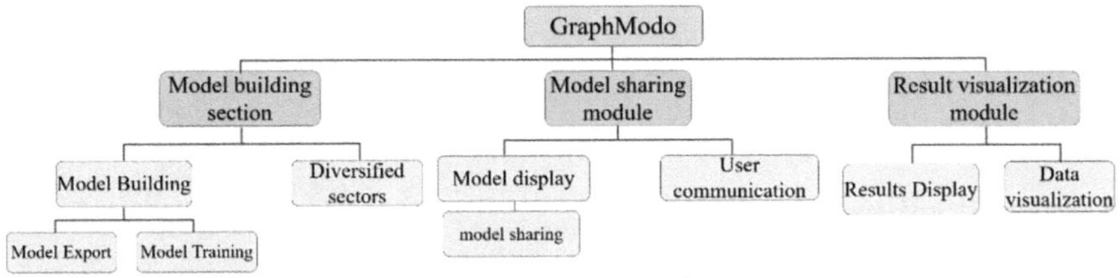

Fig. 2. Overall Functionality of GraphModo System

1. **Model Construction**: GraphModo features an innovative graphical user interface that allows users to intuitively drag and drop node. The system also includes a rich library of classic models that users can fine-tune or innovate upon.
2. **Model Sharing**: GraphModo establishes a model sharing platform that provides users with a wealth of model resources, helping them quickly find solutions that meet their specific needs.
3. **Result Visualization**: GraphModo comes equipped with powerful visualization tools that present data from model training and evaluation processes in the form of charts, curves, and graphs.

3.2 Data Structure Design

To visualize the configuration of the model in the model building interface, take the YOLOv8 model as an example, we designed a data structure that captures the key attributes of each component. The data structure is illustrated in Fig. 3.

The data structure consists of six key fields: The 'Name' field represents the module name, such as Conv2, SPPF, or Concat; the 'Form' field contains subfields 'details' and 'createTime', recording detailed information about the module and the creation time; the 'Arg' field stores the configuration parameters of the module, such as kernel size, stride, and padding for convolutional layers; the 'nodeStyle' field defines the visual style of the module in the Frontend interface, including color, shape, and border; the 'iconStyle' field specifies the icon's style for the module, including size, color, and decoration; and the 'Pos' field indicates whether the module is part of the main network or the detection head, which is crucial for correct layout and display of the module.

This design ensures the accurate representation and interactivity of the model on the Frontend page, facilitating users in building and understanding deep learning models.

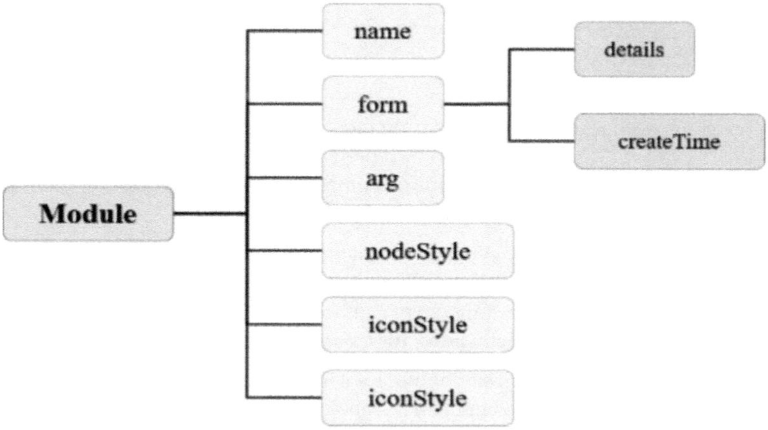

Fig. 3. Module Representation Data Structure

3.3 Low-Code Visual Model Design

GraphModo offers a low-code visual model design platform that allows users to easily build and optimize models through simple operations. The frontend VUE component relationship diagram of GraphModo is shown in Fig. 4.

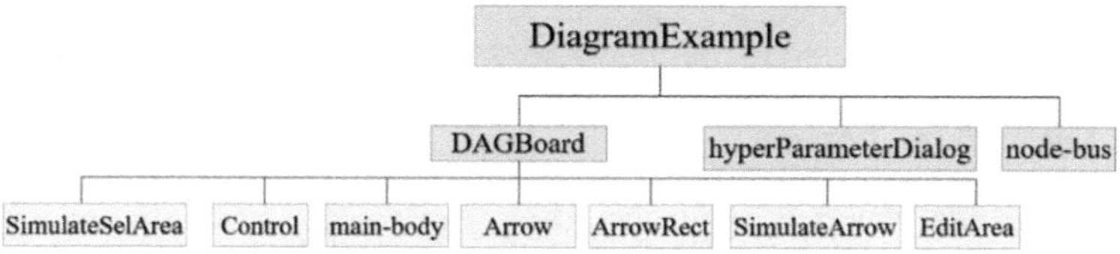

Fig. 4. Visual Model Building Features

To lower the technical barrier for designing complex deep learning models, GraphModo offers a low-code visual approach, enabling users to build and optimize models through simple operations. The DiagramExample frontend consists of three main components: DAGBoard for model construction and module management, HyperParameterDialog for hyperparameter configuration, and node-bus for simulating node placement.The module construction principle is shown in Table 2.

In the constructed model, module relationships are represented using edges to illustrate dependencies, with target data generated at the output end. The collection of operations F is viewed as a data transformation process where each operation f ∈ F functions to accept input data, process it according to transformation logic, and convert it into a specific output format. During data processing, an operation sequence S = $\{f_1, f_2, \ldots,$ fn$\}$ is constructed, where n represents the number of operations, and each operator fi receives input from fi-1 and passes the computed output to the input of fi+1.

Table 2. Principle of drag and drop model construction

step	Principle
Node represents component	In drag and drop modeling, each component or module can be represented by a node in the graph. These nodes can represent data processing, computation, and visualization Equivalent components of different types
Edge represents relationship	The edges in the diagram can represent the connections or dependencies between components. For example, the output of one component can be used as the input of another component, which allows for Connect them with an edge
Topological sorting	We can use topological sorting to determine the execution order of components, Topology sorting can ensure that the correct order is followed without cyclic dependencies Execute each component
Drag and drop interaction	A graphical model can be can structed by dragging nodes and edges, When a user drags a node, the system allows them to connect to other nodes and automatically adjust the structure of the graph based on the connection relationship
Execution model	And once the user completes the model construction, the system can execute each component in the correct order based on the topology sorting results, thereby achieving the functionality of the entire model

3.4 File Export Functionality

Graphmodo has file export capabilities. Users can select and configure the required hyperparameters through an intuitive graphical user interface. GraphModo offers a total of 31 predefined hyperparameter options for users to choose from. Some of these hyperparameter options are illustrated in Table 3.

Table 3. Selection of hyperparameters

Hyperparameter name	effect
Scale	Used for image scaling
Shear	Used for image cropping
Perspective	Used for image perspective transformation
Flipud	Used for inverting images up and down
Fliplr	Used for flipping images left and right
Mosaic	Used for image mosaic processing
Mixup	Used for image blending and enhancement

The data flow for the file export process is illustrated in Fig. 5.

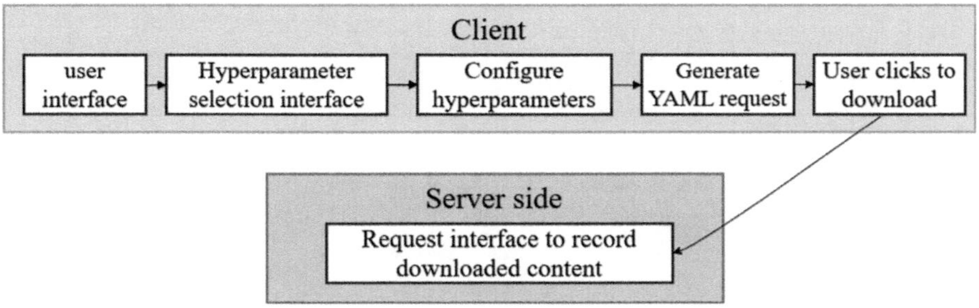

Fig. 5. Data Flow Diagram for File Export

To facilitate management and identification, users can assign a name to the exported YAML file. In the export interface, after the user inputs the file name, the system uses this name for the exported file. Once users have selected the hyperparameters and named the file, clicking the export button automatically generates a YAML file containing the chosen hyperparameter configurations and presents a download prompt. Users can then select the download path and save location.

3.5 Sharing Platform Features

GraphModo introduces a sharing feature that streamlines the model file management process. The detailed data flow diagram is shown in Fig. 6. This feature includes the following steps:

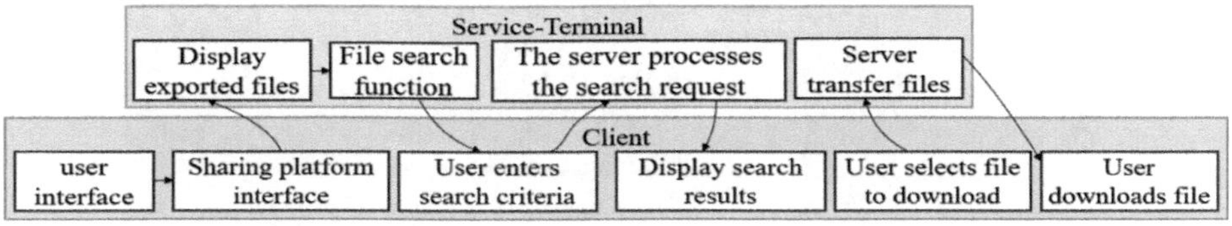

Fig. 6. Data Flow Diagram for the Sharing Platform

- File Display: The platform displays exported model files, showing information such as the download time, the downloader's name, the file name, and hyperparameters.
- Search Functionality: Users can search for files by file name or hyperparameters to quickly locate the files they need, thereby facilitating the sharing process.
- Download Functionality: Users can download the desired files either from the search results or directly from the displayed list.

The sharing platform interface is shown in Fig. 7:

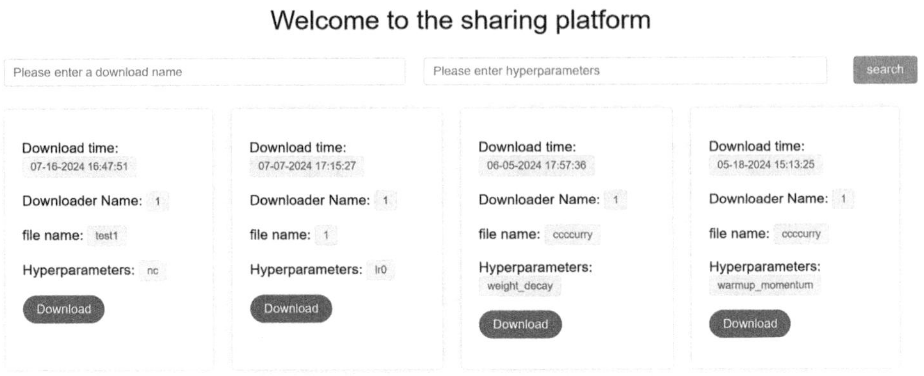

Fig. 7. The Sharing Platform Interface

4 Display of the System

This section introduces the frontend design of the GraphModo system. The system's frontend interface mainly includes three components: 1) the Construction Center Interface, 2) the Training Results Visualization Interface.

After logging in, users access the Construction Center Interface with a top menu bar for Message Center, Visualization Center, and Nodes. The central canvas allows for node management and displays dependencies between operations. The Construction Center Interface shows in Fig. 8.

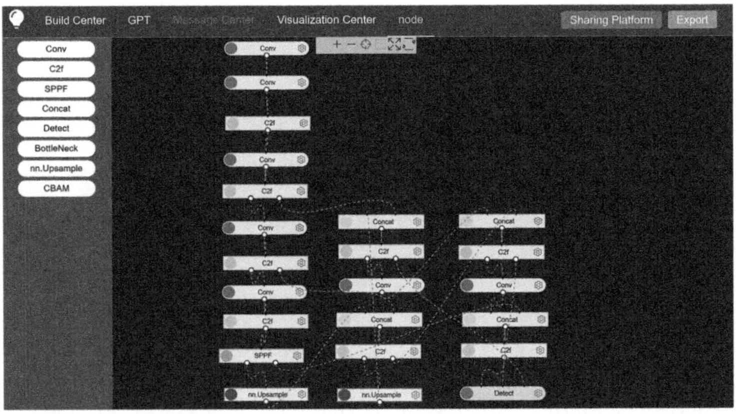

Fig. 8. Construction Center Interface

Users can drag nodes from the menu bar to the canvas to build models. Clicking the settings icon on a node opens an editing panel for entering parameters like output channels or kernel size. The "Export" button in the top-right corner allows users to specify hyperparameters for the model nodes. Fig. 9 illustrates the advanced tools available in the GraphModo system for visualizing training results.

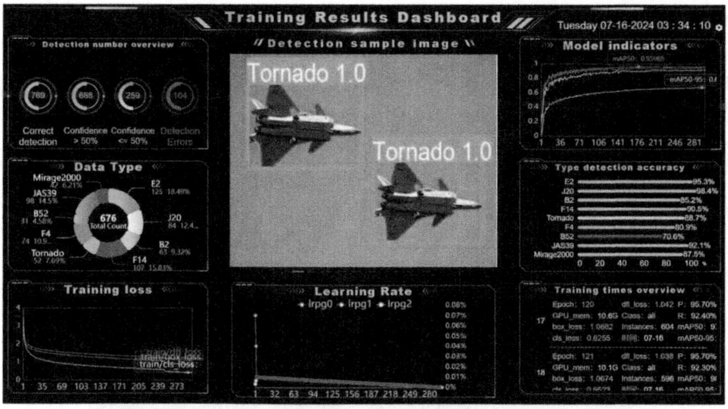

Fig. 9. Training Results Visualization Platform

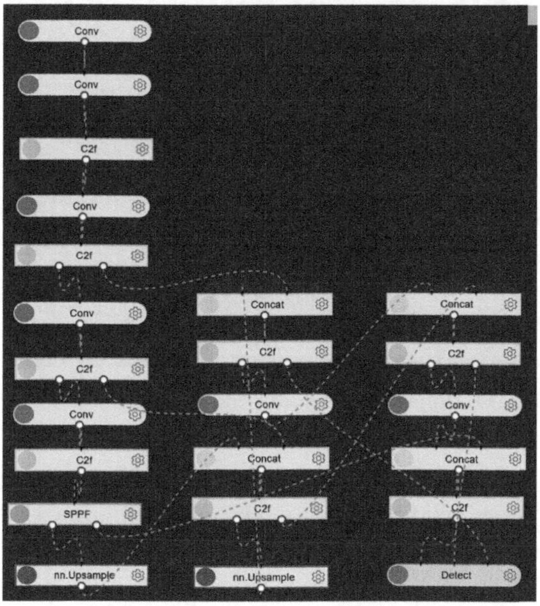

Fig. 10. YOLOv8 Model Network Structure

After completing the model construction, users can fine-tune their models and monitor the training progress in real-time. The platform provides various visualization charts, such as line graphs and scatter plots, to track important metrics including result confidence, learning rate, training loss, and model performance indicators.

5 Model Construction

Using YOLOv8 as an example, we constructed a conventional YOLOv8 model with GraphModo, and the network structure of the model is shown in the Fig. 10.

To improve the YOLOv8 model, we added a CBAM attention mechanism after the Conv modules of the three C2f modules in the Backbone. The network structure after this improvement is shown in the Fig. 11.

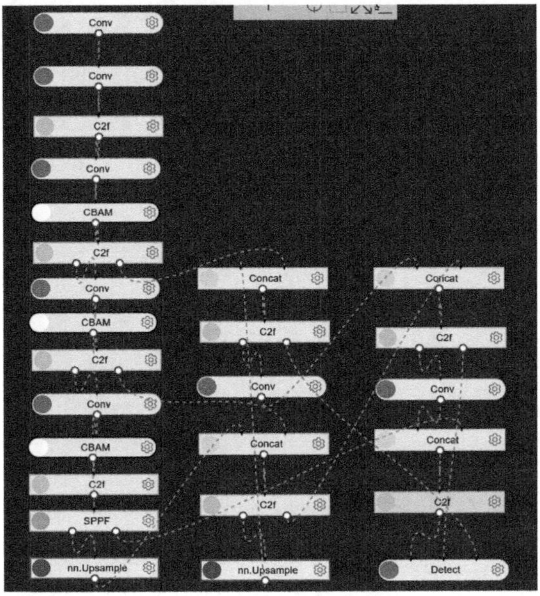

Fig. 11. Improving YOLOv8 Model Network Structure

After clicking the yaml file export button, we compared the YAML files of the Backbone area of the original YOLOv8 model and the improved YOLOv8 model, as shown in the Fig. 12. The results show that the CBAM was successfully added to the desired position, and the YAML file was updated and adjusted accordingly.

This improvement not only demonstrates the flexibility and ease of use of GraphModo in model construction but also proves that it can quickly and accurately customize and optimize models in practical operations.

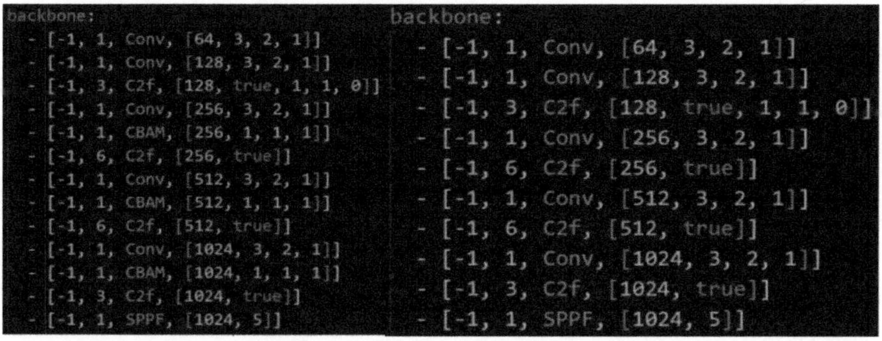

Fig. 12. Comparative Analysis of Improved and Original Model YAML Files

6 Conclusions

The test team of this project conducted a comprehensive and detailed assessment of the GraphModo system. After more than six months of in-depth research and persistent efforts, the system has undergone multiple iterations and performance optimizations, resulting in significant improvements in functionality and performance. User feedback has been positive, and the interface design is user-friendly, greatly enhancing the user experience.

GraphModo integrates the visualization and construction of deep learning models, featuring an intuitive interface, flexible configuration options, convenient export functionality, powerful visualization tools, and a collaborative sharing platform. The system continuously expands its model library, aiming to incorporate various deep learning models and drive technological advancement.

Despite these improvements, some areas require further enhancement. Performance bottlenecks may arise when handling large datasets, necessitating efficiency and response speed improvements. Compatibility issues with certain hardware or software environments need further optimization. Additionally, data privacy and security mechanisms must be more robust. Addressing these shortcomings will enhance GraphModo's competitiveness, making it a more complete and popular system for deep learning model visualization and construction.

As an advanced and innovative model visualization and construction system, GraphModo holds significant development potential. With the rapid advancement of deep learning technology, GraphModo is poised to become a leading platform, offering intuitive tools and a comprehensive model library to facilitate the widespread adoption of deep learning technology. It will provide customized solutions for various industries, aiding in their digital transformation, and serve as a platform for exchange and collaboration in the deep learning field. By fostering an active community and incorporating continuous user feedback, GraphModo aims to drive technological progress and application. Dedicated to connecting deep learning practitioners worldwide, it promotes knowledge sharing and collaborative exchange.

Acknowledgment. The authors express their sincere gratitude for the support received from the National College Students' Innovation and Entrepreneurship Training Program (Project No. 202313684005X) and the Guangdong Province College Students' Innovation and Entrepreneurship Training Program (Project No. S202413684014).

References

1. Liu, X., Li, L., Zhang, J.: A comprehensive review on experiment tracking tools for machine learning research. IEEE Access **10**, 120579–120592 (2022). https://doi.org/10.1109/ACCESS.2022.3232235
2. McGhee, J.G., Clark, P., Larkin, A.R.: Paperspace gradient: a managed platform for building and scaling machine learning workflows. In: Proceedings of the 2022 International Conference on Machine Learning (ICML), pp. 1–10 (2022). https://doi.org/10.48550/arXiv.2204.11288
3. Ohmura, H., Nakashima, H., Okamoto, T.: A Survey on Deep Learning Tools for Visualization and Experiment Management. ArXiv Preprint, 2002.12663. https://arxiv.org/abs/2002.12663. (2020)
4. Tang, S.Z., Zhang, Y.F.: DragDL: an easy-to-use deep learning model visualization and construction system. Comput. Sci. **48**(08), 220–225 (2021). https://doi.org/10.3969/j.issn.1002-137X.2021.08.026
5. Ma, Y.J., Liu, H., Zhou, T., Chen, W., Yu, X., Lin, C.: PaddlePaddle: an open-source deep learning platform from industry practice. Data Comput. Dev. Front. **1**(05), 105–115 (2019). CNKI:CNKI:SUN:KYXH.0.2019-05-011

Author Index

G. Jeon and X. Kong (Eds.): ACIIBD 2024, LNICST 616, pp. 177–178, 2025.
https://doi.org/10.1007/978-3-031-91853-7

GPSR Compliance

The European Union's (EU) General Product Safety Regulation (GPSR) is a set of rules that requires consumer products to be safe and our obligations to ensure this.

If you have any concerns about our products, you can contact us on ProductSafety@springernature.com

In case Publisher is established outside the EU, the EU authorized representative is:

Springer Nature Customer Service Center GmbH
Europaplatz 3
69115 Heidelberg, Germany

Batch number: 08544862

Printed by Printforce, the Netherlands